Moral Judgments as Educated Intuitions

Moral Judgments as Educated Intuitions

Hanno Sauer

The MIT Press
Cambridge, Massachusetts
London, England

This book was set in Stone Sans and Stone Serif by Toppan Best-set Premedia Limited.

Library of Congress Cataloging-in-Publication Data

Names: Sauer, Hanno, author.
Title: Moral judgments as educated intuitions / Hanno Sauer.
Description: Cambridge, MA : MIT Press, 2017. | Includes bibliographical references and index.
Identifiers: LCCN 2016023438 | ISBN 9780262035606 (hardcover : alk. paper),
ISBN 9780262546706 (paperback)
Subjects: LCSH: Judgment (Ethics) | Ethics--Psychological aspects. | Emotions. | Ethical intuitionism.
Classification: LCC BJ1408.5 .S38 2017 | DDC 170/.42--dc23 LC record available at https://lccn.loc.gov/2016023438

For Mina

Contents

Acknowledgments

This book grew out of a dissertation I wrote between 2009 and 2013 at the Universities of Leiden and Groningen. I would have neither started nor finished this project had it not been for my supervisor, Pauline Kleingeld, and her encouraging feedback, moral support, and friendly tenacity. When I arrived in the Netherlands in May 2009, she had done most of the important work already; I only had to fill in the gaps. I hope I did not depart too much from her Kantian proclivities, but I am confident that she will tolerate whatever heresy I have come up with.

I also wish to thank NWO (the Netherlands Organisation for Scientific Research) for their generous funding of the project "Morality beyond Illusions? Re-assessing the Philosophical Implications of Empirical Studies of Moral Agency," of which I was part. The time I spent working on our project in Leiden was one of the most exciting and stimulating periods of my life. I want to thank Markus Schlosser, from whom I learned more than he knows, and Tom Bates, from whom I learned less than he thinks.

I am indebted to the following whose support, feedback, and suggestions make them partly responsible for the mistakes that can be found on the pages below: Joel Anderson, Mark Alfano, Alexander Bagattini, Bert Bos, Wout Cornelissen, Sabine Döring, Julia Driver, Fabian Eck, Yvonne van Eijk, Thomas Fossen, Hedwig Gaasterland, Victor Gijsbers, Sean Gould, Carolyn de Greef, Thomas Grundmann, Frank Hindriks, Daniel Jacobson, Lies Klumper, Antti Kauppinen, Jouni Kuukkanen, Joseph Heath, Anthonie Meijers, Tiffany Meshkat, Marijana Milosavljevic-Vujosevic, Sven Nyholm, James Pearson, Jeroen van Rijen, Scott Robbins, Philip Robichaud, Sabine Roeser, Herman Siemens, Bart Streumer, Alan Thomas, Peter Timmerman, Bruno Verbeek, Jonathan Webber, and all the others whom I've forgotten. I do not thank my landlord.

I received a lot of helpful comments from audiences at various conferences and colloquia in Amsterdam, Barchem, Cardiff, Eindhoven, Groningen, Leiden, Milan, Rotterdam, Tilburg, Toronto, Tübingen, Utrecht, and Zürich. I also wish to thank my students at Leiden and Tilburg, from whom I have learned much. I would also like to thank Philip Laughlin for his generous and professional support, and three anonymous reviewers at the MIT Press for their constructive and sympathetic comments. I am grateful to Matthias Warkus for preparing the index.

Special thanks go to Jeanette Kennett, Maureen Sie, and Jesse Prinz for reading the whole manuscript, for agreeing to serve on my committee, and for the valuable advice they have given. I am grateful to Jesse Prinz for hosting me at the CUNY Graduate Center and the time and effort he put into reading and commenting on my biweekly produce, to Axel Honneth, for hosting me at Columbia University, and to both for giving me a reason to spend three months in New York City.

I owe a lot to my academic teachers at Marburg and Frankfurt. Peter Janich, Udo Tietz, Marcus Willaschek, and my MA supervisors Axel Honneth and Rainer Forst all profoundly shaped my philosophical views. I see their influence on every page of this book, even if they may not.

I also feel the need to thank many people I may have never met and perhaps never will, in particular the anonymous referees from various journals whose insightful comments I benefited from. This book is much better because of them.

I have reused material from previously published articles. Chapter 1 is based on "Social Intuitionism and the Psychology of Moral Reasoning," *Philosophy Compass* 6 (10): 708–721; chapter 2 draws on material from "Educated Intuitions. Automaticity and Rationality in Moral Judgment," *Philosophical Explorations* 15 (3): 255–275; chapter 4 is based on "The Wrong Kind of Mistake: A Problem for Robust Sentimentalism about Moral Judgment," *Journal of Value Inquiry* 48 (2): 247–269; chapters 5 and 6 reuse portions of my "Psychopaths and Filthy Desks. Are Emotions Necessary and Sufficient for Moral Judgment?" *Ethical Theory and Moral Practice* 15 (1): 95–115; and finally, chapter 7 is based on "Morally Irrelevant Factors: What's Left of the Dual-process Model of Moral Cognition?" *Philosophical Psychology* 25 (6): 783–811. I am grateful to the editors and publishers of the above journals for granting me permission to use this material.

I want to thank my family and friends. My parents always allowed my interests to thrive; I thank them for making me a philosopher and, *a fortiori*, an utterly useless member of society. I want to thank my brothers, David and Moritz, for being there and supporting me. I thank my parents- and siblings-in-law, Nono, Aida, Benko and Vera. Most of all, I want to thank my wife, to whom this book is dedicated. Mina—*wir beide*.

Introduction

1 The Fork

In eleventh-century Venice, the doge married a Greek princess with peculiar habits: when she ate, she used a small golden tool with two tines to carry her food from the plate to her mouth. This tool—nowadays referred to as a "fork"—was unheard of at the time, and when word got around about her deviant behavior, it sparked a considerable scandal among the honest Venetians. The dogaressa died of a vicious disease shortly afterward; meanwhile, the citizens of medieval Venice unanimously greeted the poor princess's fate with approval, seeing her early death as divine punishment for the sinful delicacy of her manners.

Why did they care so much about the fork? The Venetians did not seem to be merely puzzled by the princess's unusual behavior, but genuinely *outraged* by her precarious diligence. They were not annoyed by what she had done, nor did they deem it imprudent; they did not, as far as we know, think it harmful to themselves or any other person. And they certainly were not afraid of it. They simply found the way she deviated from the norm to be appalling and offensive. In short: they thought it was *morally objectionable*.[1]

Morality is no joke. People kill and die for it. It pervades our lives—it shapes the way we think, structures the way we feel, and alters the way we behave. But it does so subtly and quietly. No overt force is needed for moral norms and values to reach through our intentions and exert their power upon our actions, and we need not be coerced into thinking in moral terms about ourselves and others. It almost comes naturally to us, and colors the way we perceive the world in the most fundamental way: once we have been introduced into a moral code, complying with its awkward demands seems as inescapable to us as the laws of physics. Moral norms issue commands to do some things and to omit others, and we obey. Oftentimes, the question whether we should or should not behave as morality requires does not even cross our minds.

Why is that? In large part, this is due to the extent to which morality recruits the most basic elements of the human mind: the emotions. This book is about morality, emotion, and what the two have to do with each other. Some theorists argue that morality is all about emotion, and little or nothing else. That is how morality "binds and blinds" (Haidt 2012). Moral emotions make people do good and bad things, and they give them the ability to assess agents and actions in terms of good and bad in the first place. They obstruct and facilitate people's sense of right and wrong. Emotions make them care about some things and dismiss others as not worth caring about. They determine who they like, what political views they have, who they want to hang out with after work, which films they get to see and which tools they use to carry their food from plate to mouth. But emotions are not the only things that play a role in morality. Or are they?

2 What's Reason Got to Do with It?

We do not wish death upon those who use forks—not anymore, at least. Norms have changed, and nowadays, *not* using a fork has become extravagant, and people who refuse to follow the crowd in this respect don't often get invited a second time. Other things have changed, too: we do not eat from the same bowl or spit under the table anymore. And one might think that all of this makes sense: it is risky and unhygienic to do any of the above, and it is good that we have abandoned those atavistic habits.

But, as Nobert Elias argued in his monumental *Civilizing Process* (1994), this change in our emotional attitudes toward forks and shared dishes had nothing whatsoever to do with rational insight. A quick glance at how people actually justified their refined sense of table manners reveals that at the time, people could not have cared less about the harms that were brought about by spreading germs, which, needless to say, they did not even know existed. Instead, people were content with saying that it wasn't "polite" to blow one's nose on the tablecloth; that it wasn't "courteous" to scratch one's back with the communal spoon; and that it would be considered "offensive" or "embarrassing" to fart just when the pheasant was served.

Rational justifications in terms of harm and hygiene merely hijacked those changes in their feelings after they had already occurred. Reasoning and argument did not bring them about. They merely put a stamp of approval on what had already been decided by the evolution of people's emotional sensitivity.

This phylogenetic account is, in a nutshell, exactly what many moral psychologists believe applies to the ontogenesis of moral judgment and its relationship to moral reasoning. Individual people form moral judgments and societies at large live by certain moral codes. But which moral beliefs people hold and which system of norms and values is in play in a given social context is not, these psychologists argue, a matter of reason and genuine insight. Rather, it is a matter of feeling and intuitive gut reactions. Sophisticated justifications for why some things are wrong and some right are invented after the fact, in order to rationalize the moral norms and values one already happens to have. This suggests that reasoning plays no role whatsoever in which moral beliefs people hold. Or does it?

3 Two Challenges

So far, I have preliminarily introduced two ideas. The first one was that morality has everything to do with emotion. The second was that it has nothing to do with reason.

An influential philosophical account of the nature of moral judgment, however, rejects both these ideas. This account holds that emotions and feelings play little (or no important) role for moral judgment, and that moral cognition is thoroughly based on reason and thinking. Traditionally, this type of account has been referred to as *rationalism* about moral judgment. Recently, however, rationalism—as I will mostly refer to it in what follows—has come under serious attack. This attack is largely based on two challenges, which bring to bear novel empirical data and cutting edge experimental methods to revive the above two ideas. I will refer to these as the *emotionist* challenge and the *anti-rationalist* challenge, respectively.

(1) *The emotionist challenge.* Call the claim that emotions are, in some sense yet to be specified, essential for morality and moral judgment *emotionism about morality*. (The traditional name for this position is sentimentalism, but I will stick to the former because of its closer connection to contemporary debates in moral philosophy and psychology.) This

challenge comes in a variety of forms: the Humean, internalist version of this challenge has it that due to the conceptual connection between moral judgment and motivation, and the fact that only noncognitive states such as desires and emotions have motivational power, moral judgments must be noncognitive states as well. Therefore, this argument goes, moral judgments are not grounded in reason. The expressivist version of this challenge has it that moral judgments do not purport to describe facts, but serve to express certain normative attitudes such as preferences, feelings, or desires. Therefore, this argument goes, moral judgments are not grounded in reason.

In what follows, I will not delve into the details of these arguments, but focus on a more recent version of the emotionist challenge. The most significant contemporary attempt at undermining the rational foundation of morality is based primarily on findings in the empirical psychology of moral judgment. This is not to say that the aforementioned, more traditional versions of the emotionist challenge have no empirical implications, or are not susceptible to an assessment in light of empirical facts. They are indeed—but their case is not based on empirical considerations right from the start; their validity does not depend, in the same direct way, on experimental results and empirical research; and they do not make straightforward empirically testable predictions.

This, however, is exactly what the emotionist challenge I am concerned with in this book aims to do. Take this example: there is evidence that people who suffer from impaired emotion have trouble developing full moral competence, and end up making poor moral judgments. Or this: many studies show that by manipulating people's emotions, one can change their moral beliefs. In what follows, I will explain this evidence in great detail, and offer a novel interpretation of its implications; for now, let it suffice to say that on the basis of evidence for the emotional foundation of moral judgment the emotionist challenge aims to establish empirically—although, importantly, not *only* empirically—that rationalism about moral judgment is false.

(2) *The anti-rationalist challenge.* Call the claim that reason and reasoning play no significant role for morality and moral judgment *anti-rationalism about morality*.

Traditionally, this second challenge was often thought to be redundant, as it seemed to follow directly from the emotionist challenge. If emotion and reason are incompatible with each other, and moral judgment is based on emotion, then it could not possibly be based on reason. But recent empirical evidence suggests that the two challenges are distinct, and that the influence—or lack thereof—of actual *reasoning* on subjects' moral beliefs can be studied directly. People frequently arrive at their moral verdicts so quickly that it would be impossible for them to have been preceded by any type of rational deliberation. Moreover, there are studies suggesting that people's moral reasoning is thoroughly *epiphenomenal*: people arrive at their judgments intuitively and automatically; episodes of conscious reasoning merely rationalize these intuitions *after the fact*. Again, I will explain these results in great detail below; for now, I will just remark that on the basis of evidence against the rational foundation of moral judgment the anti-rationalist challenge aims to establish empirically—although, at least as importantly as in the first case, not *only* empirically—that rationalism about moral judgment is false.

It is important to note that, although there is a certain elective affinity between these two challenges, they are logically independent from each other. One could be an emotionist about morality, and hold that moral values are grounded in emotion, while thinking that reasoning is often required when it comes to figuring out whether someone's values are violated or instantiated in a given case. On the other hand, one could be an anti-rationalist, and think that moral reasoning is a *post hoc* affair, without having to agree that the norms and values we latch onto with our automatic, reasoning-independent intuitions depend on our feelings.

I aim to defend the rationalist account of the psychology of moral judgment against these two challenges. I will argue, first, that the anti-rationalist challenge can be *met*, as it can be shown empirically that reasoning does play a crucial role in moral judgment. On the other hand, what can be learned from the evidence in support of the anti-rationalist is that one should be wary not to overestimate the importance of moral reasoning and be prepared for reason to exert its influence upon moral judgment in unexpected ways. I shall explain later what these unexpected ways are, and how one can meet the anti-rationalist challenge while retaining its critical potential. Second, I will argue that the emotionist challenge does not have to be met, because it can be *dissolved*. The idea that the importance of emotions for moral judgment threatens the prospects of rationalism rests on a series of faulty conceptual presuppositions, virtually all of which can be debunked. A comprehensive account of moral judgment specifies why and where emotions come into play in how people arrive at their moral beliefs, and shows how reasoning and emotion interact in the formation of moral judgments to constitute full-blown moral agents. To put it into a slogan: the anti-rationalist challenge is a real challenge, but its core tenets aren't true; the emotionist's claims are true, but do not pose a real challenge.

4 Feeling and Thinking: A (Very) Brief History of Recent Moral Psychology

Methodologically speaking, what follows can be seen as an exercise in empirical philosophy (Prinz 2008). Empirical philosophy does not abandon all interest in traditional philosophy's core questions, but it hopes to make progress on them by getting down to business with Hume's promise to "introduce the experimental method into the moral sciences." It does not give up on the analytic style either, but hopes to replace the frictionless spinning in the void of mere conceptual analysis with a conceptually rigorous and empirically informed approach.

If one had to choose the *annus mirabilis* of the naturalistic turn in metaethics (Prinz 2015), 2001 would be the obvious choice. Both Jonathan Haidt's groundbreaking paper, "The Emotional Dog and Its Rational Tail," as well as Joshua Greene's first study on the neural basis of moral cognition appeared in this year. John Doris's situationist manifesto *Lack of Character* was published only a year later. Ever since then, the trend has continued: nowadays, empirical philosophers routinely draw on a wealth of research from social psychology, cognitive neuroscience, anthropology, sociology, and history in order to buttress their metaethical claims. From the very beginning, the default view regarding the nature of moral judgment in empirically informed metaethics has been emotionism. Because of that, my argument will, to a large extent, consist in a defense against the emotionist and the anti-rationalist challenge to the rationalist position. The *educated intuitions* account I aim to outline is supposed to develop an all of a piece response to both of them.

(1) *Social intuitionism and dual process theory.* Haidt's Social Intuitionist model holds that moral judgments are produced by quick, emotionally charged intuitions, and that conscious reasoning provides *post hoc* rationalizations in defense of those. Haidt's claims sparked nothing less than a

revolution in moral psychological research, which had grown increasingly discontent with Lawrence Kohlberg's (1969) intellectualist emphasis on reasoning and moral progress. Greene's dual process model of moral cognition, on the other hand, did not take sides in the rationalism vs. sentimentalism debate. Unlike the more straightforwardly Humean theories by Shaun Nichols, Haidt, or Jesse Prinz, it holds that the distinction between emotion and reason cannot be mapped on the distinction between moral and nonmoral judgment, but on the distinction between different subsets of moral judgment—namely, paradigmatically deontological and paradigmatically consequentialist moral judgments.

(2) *Universal moral grammar theory*. Soon after the empirical case for the automaticity of a large amount of moral judgment was first articulated, another approach stepped on the scene, which, though it applauded Haidt's and Greene's suspicious attitude toward the importance of conscious reasoning, did not go down the same anti-rationalist path. John Rawls (1971, 47) was the first to suggest a "linguistic analogy" between ordinary structures of grammar, as described by Noam Chomsky (2009) and others, and the structure of moral cognition. Today, John Mikhail (2007) and Marc D. Hauser (2006) aim to explain certain central moral principles and distinctions—such as the doctrine of double effect—in terms of a Universal Moral Grammar. We have an innate ability, this theory says, to produce structural descriptions of actions in terms of various morally relevant categories, and to pass moral verdicts on the basis of those descriptions.

(3) *Sentimental rules and constructive sentimentalism*. Shaun Nichols's (2004) sentimental rules account was the first one to make an attempt at developing a theoretically comprehensive, philosophically interesting and empirically informed account of what he referred to as "core moral judgment"—judgments about behavioral transgressions of nonconventional, emotion-backed norms. To make such judgments, Nichols argued, one needs two things: a normative theory that specifies a set of rules prohibiting harm, and a set of emotional responses attuned to this normative theory. Nichols's pioneering account was the first to apply the latest empirical methods to the philosophical study of virtually all aspects of moral judgment—their psychological basis, their motivational role, and their origins. Jesse Prinz's (2006, 2007) constructive sentimentalism is bold and simple, and follows Nichols's example in most methodological respects—as in Hume's case, its sophistication lies in the defense. The theory says that

moral judgments refer to moral properties, and that both judgments and their truth-makers are emotional dispositions. Moral judgments are emotional attitudes; moral properties are corresponding and response dependent. This leads to a theory that is openly relativist and subjectivist. In many ways, Prinz's theory can be seen as the most radical and unapologetic statement of naturalized sentimentalism.

Of these five theories, only two—Haidt's and Prinz's—explicitly set up their accounts as challenges to rationalism. Although Nichols argued for a sentimentalist position, he mostly seemed preoccupied with showing that the constraints *neo*-sentimentalists impose on the notion of moral judgment are, from a naturalistic point of view, too demanding (Nichols 2008). Greene's theory did not attack rationalism about moral judgment per se, but leveled its charges against rationalism about *deontological* moral judgment. Universal Moral Grammar theorists were mostly interested in the nature/nurture aspect of the debate, arguing that many core principles of moral judgment are universally shared and thus perhaps innate. And while they did agree that those innate principles often remain inaccessible to conscious reasoning, they also seemed to think that our judgments are both guided by these principles and that these principles are rationally acceptable simply in virtue of the fact that, being hardwired features of our cognition, they are without viable alternatives. In what follows, I shall therefore treat the emotionist challenge exemplified by Prinz and the anti-rationalist challenge exemplified by Haidt as rationalism's main opponents: the first for arguing that, regardless of the role conscious reasoning may play when people have to figure out whether a particular norm or value applies, moral norms and values are constituted by emotional dispositions; the second for arguing that reasoning plays no *real* role for moral judgment at all. In the first case, I shall explain why the emotional foundation of moral judgment, once it is properly understood, poses no threat whatsoever to rationalist accounts of moral cognition. In the second case, I shall argue that Haidt grossly misrepresents the relationship between automatic intuitions and conscious reasoning. Once this relation is properly understood, it becomes apparent that in the light of the available evidence, a moderate form of rationalism offers the best overall account of moral judgment.[2]

The educated intuitions account aims to overcome as many of the problems that plague the aforementioned theories as possible. It explains

(1) *contra* the sentimental rules account, how exactly normative theories and emotional reactions interact to produce moral judgments;
(2) *contra* constructive sentimentalism, how to incorporate the importance of emotion for moral judgments, while making understandable the possibility of genuine moral error, and how to do justice to the various conceptual constraints on the notion of moral judgment;
(3) *contra* social intuitionism, how to draw the line between benign *post hoc* reasoning and confabulatory after the fact rationalizations, why rationality does not require conscious reasoning, and why moral reasoning is an inevitably social enterprise;
(4) *contra* dual process theory, that regarding how subjects make moral judgments, there is no clear cut distinction between dissociable mental subsystems, corresponding normative statuses and conflicting moral theories; and
(5) *contra* Universal Moral Grammar theory, that the foundations of moral judgment do not lie in an eternally fixed Universal Moral Grammar that allows only for very little education and improvement and concedes only a marginal role, if any, to the acquisition of new moral norms and principles.

To be sure, there are some advocates of the rationalist cause. My aim is to follow up on and complement these defenses, and to supply a systematically worked out framework describing the respective roles that reason, intuition, and emotion play for human moral thinking. Here are some examples of rationalist responses to the emotionist dominance in recent empirical metaethics, and the main ideas they have drawn upon:

(1) *Conceptual constraints*. Some have tried to respond to the emotionist challenge by making a conceptual move: our concept of moral judgment, this argument goes, is such that it requires at least some connection to reasoning and justification (Kennett and Fine 2009). Any theory of moral judgment that ignores or leaves this connection unaccounted for fails at the most fundamental level of identifying the correct *explanandum* in the first place. Without a sufficiently rich grasp on what moral judgments are supposed to be, even the best empirical evidence necessarily misses its target, and cannot be brought to bear on the question of interest. A simple reduction of moral judgments to emotional attitudes faces exactly this problem.

(2) *Moral agency*. A second type of rationalist response emphasizes the role of moral agency for moral cognition (Gerrans and Kennett 2010). According to this response, emotionist as well as intuitionist theories suffer from a distinctive form of myopia, and misconstrue moral judgments as time-slice flashes of emotional approval or disapproval, rather than parts of a person's temporally extended moral agency.

(3) *Intuitive principles*. Most traditional and modern forms of moral intuitionism have no interest in whether their moral epistemology or metaphysics are empirically feasible. But this need not be so: there are ways in which one can understand the immediate and perceptual features of moral intuition in a way that makes intuition understandable from a naturalistic point of view while retaining intuitionism's rationalist commitments. In this spirit, Terry Horgan and Mark Timmons (2007) plead for a form of morphological rationalism that aims to detect traces of abstract moral principles in people's psychologically immediate intuitive responses.

(4) *Reflection*. Rationalists trust reflection. But although this trust should not be overblown (Bortolotti 2011; Arpaly and Schroeder 2012), reflection does sometimes perform important functions. In the case of moral cognition, this has been emphasized by, among others, Karen Jones (2006), Albert Musschenga (2008), Jeanette Kennett and Cordelia Fine (Kennett and Fine 2009), Jillian Craigie (2011), and Lorraine Besser-Jones (2012): they all insist on the role that conscious deliberation often plays in moral judgment, and urge us not to underestimate the power of rational thinking. Emotion and intuition yield quick responses that are often reliable, and do not draw on expensive cognitive resources. At the same time, they can be crude, selective, and can easily mislead. Reflection can redirect one's attention to previously overlooked morally relevant features of situations, extraneous influences on one's judgments and more abstract considerations that sometimes escape subjects' intuitive awareness.

The educated intuitions account aims to synthesize the insights that can be found in the above approaches, and forge them into a comprehensive rationalist theory of moral judgment that is supported by the empirical evidence for the influence of reason *as well as emotion* on moral judgment, acknowledges the crucial role automatic intuitions play in moral cognition, and offers an account of the justificatory status of moral intuitions and the role they play within the social practice of moral reasoning.

5 Educated Intuitions: An Overview

This is a theory about how moral judgment works. More precisely, it is a theory about whether the way moral judgment works makes it legitimate to describe it as an exercise of reason. In addressing this question, I shall argue, most authors have ended up with anti-rationalist versions of emotionism or intuitionism because they have looked too closely at what is going on in people's minds: it is true that *particular* moral judgments made on *particular* occasions are typically made on the basis of emotionally charged automatic intuitions. When subjects make individual moral judgments within a confined point in time, the visible influence of moral reasoning remains negligible.

If one adopts this approach to the study of moral judgment, one will tend to understand moral agency in terms of a continuous stream of isolated moral verdicts, none of which are arrived at via a rational route. I favor a different approach and suggest starting from the other end. The pervasive influence of reason on moral judgment becomes visible only if one does not understand moral agency as a bundle of moral judgments—which, if looked at individually, often consist in cognitively unpenetrated gut reactions—but if one describes moral judgment the other way round, as the exercise of moral agency on a particular occasion. This difference in perspective is not trivial; it opens one's eyes to the past, present, and future of morally judging subjects.

I shall argue that, although we only rarely have time to reflect about what to do on a given occasion, we did have time to acquire a repertoire of intuitions about what is morally acceptable, over the course of our moral education, which we can produce automatically without any thought. And we do have time to reflect and reason about those intuitions when we are confronted with a special reason to do so—a conversation we had, a new piece of information we gathered, an argument we embarked upon with a friend, or a moral conflict we encountered. This, I will argue, is where reason comes into play in the production of moral judgment. From the back: because reasoning figures in the acquisition, formation, and maintenance of our moral intuitions. From the front: because these moral intuitions are amenable to reflection, once the need for an intermittent episode of moral reasoning has arisen.

Moral judgments are educated and rationally amenable moral intuitions. These intuitions are typically the upshot of emotional reactions to morally salient situations—on paper or in everyday life. Intuitions are a lot like cognitive habits. They are acquired, automatized judgmental responses to scenarios that require such responses. Because these responses are mostly automatic, moral reasoning is not how subjects typically arrive at them. Reasoning is typically not what produces moral judgment; it is a *checks and balances* device that, when necessary, resolves inconsistencies, supplies our intuitive system with new information, and reflexively monitors the sustainability of our intuitive judgments. It is only called for when a given habitualized moral intuition faces rational pressure. In this case, and when things go well, reasoning either recommends giving up the intuition or provides additional grounding for it. This process then feeds back into subjects' intuitions, educating them further. In the long run, repeated cycles like this nurture individual and social moral progress.

This book has three parts. The first two each deal with one of the two challenges I have introduced above. In the first part, I explain the antirationalist challenge, excavate its central conceptual commitments, and provide a wealth of empirical evidence for the effectiveness of moral reasoning to show how it can be met. The second part is about the emotionist challenge. I explain and analyze this challenge, review the evidence there is for a tight link between emotion and moral judgment and show how, from the perspective of the educated intuitions model whose outlines are developed in the first part, a rationalist reading of this evidence can be developed. Tying the first two together, the third part turns toward the dual process model of moral cognition as a test case for the empirical viability and normative convincingness of the educated intuitions account.

Jonathan Haidt's social intuitionist model holds that moral judgments in general are based on automatic and intuitive gut reactions, and that moral reasoning in general does very little besides providing *post hoc* rationalizations for the moral intuitions people already happen to have. When people reason about their moral beliefs, they do not do so in pursuit of the truth, but in the service of social persuasion. In the first chapter, I will argue that the interpretation of the empirical evidence for the claim that moral reasoning leaves subjects' intuitions mostly untouched crucially depends on various assumptions about the nature of (conscious) reasoning that we are given little reason to accept. More precisely, I argue that this claim is based on an unjustified double standard: nonmoral and moral reasoning is virtually always a *post hoc* enterprise, and need not satisfy what I call the accessibility- and/or the causality-requirements. Social intuitionism's commitment to these two requirements renders the model unable to account for the difference between ordinary *post hoc* reasoning and genuine confabulatory rationalizations.

The following two chapters, which conclude the first part, develop a constructive response to the anti-rationalist challenge posed by the social intuitionist model. In the second chapter, I will argue that the influence of moral reasoning on people's moral intuitions need not be mediated by conscious reasoning. Moral reasoning becomes effective through the acquisition, formation, maintenance, and correction of moral intuitions. Over time, previously undertaken episodes of conscious reasoning migrate into people's intuitions, and become habitualized. I refer to this process as the *education* of our moral intuitions, show how this education might work, argue that it *does* work, and use the resulting model to distinguish between various forms of *post hoc* reasoning, some of which amount to confabulation, others to a form of making explicit the patterns of reasoning that figured in the development of the intuitions subjects rely upon in making moral judgments.

The anti-rationalist challenge to rationalism about the psychology of moral judgment is best understood as an automaticity challenge. This challenge, in turn, is based on what I call the incompatibility thesis, a claim that complements the accessibility and the causality requirement and holds that automatic judgments cannot, by their very nature, be rational. I argue that this is not correct, and indeed quite obviously so. As far as the rationality of automatic attitudes is concerned, we should rely on a parity principle for which the concept of an education of our intuitions proves to be central: if some process of judgment formation has become automatic over time (through habitualization), then there is no reason for us to think that anything about the rationality of that process has changed. I show how the concept of an education of our moral intuitions can help shed more light on the distinction between unproblematic *post hoc* reasoning and problematic episodes of confabulation. The phenomenon of moral dumbfounding can be reinterpreted with that distinction in mind. Moreover, I present a wealth of empirical evidence for the claim that our moral intuitions not only can be, but in fact are, educated and rationally amenable.

The third chapter has two aims. What still remains to be done after the main part of the anti-rationalist challenge is met is to see, first, whether the way subjects arrive at their moral judgments as described by the educated intuitions account satisfies some basic justificatory requirements. It is one thing to show that moral judgments are based on mental processes which deserve to be described as "reason," and an entirely different thing whether those processes are sufficient to render people's judgments justified. Second, a comprehensive account of moral judgment has to account for the irreducibly social dimension of moral reasoning. I will argue that the fact that moral reasoning has an important social function does not undermine the rationalist case. Cooperative moral reasoning in many ways further improves people's moral intuitions and their ability to justify their evaluative points of view. In a sense, it is fair to say that moral reasoning does not really take place within individual minds at all: as a thoroughly social enterprise, the education of the social reservoir of moral intuitions takes place *between* individual moral reasoners.

The first two chapters of the second part explain why I do not wholeheartedly embrace an emotionist account of moral judgment, despite agreeing about the tremendous influence of emotion on moral judgment. In the fourth chapter, I argue that emotionism not only makes claims about the *psychology* of moral judgment—that emotions are both necessary and sufficient for moral judgment—but also about the *metaphysics* of moral properties. Emotionism holds that both moral judgments and moral properties are constituted by emotions. This creates difficult problems for the emotionist, as it seems to make moral error impossible. Traditionally, this has been thought to be due to the fact that sentimentalism entails that moral judgments cannot be false *or true*. I argue that this is not the case for modern empirical sentimentalism. Rather, this form of sentimentalism makes moral error impossible because it entails that moral judgments are *necessarily true*. Moreover, I argue that in trying to escape this unhappy result, sentimentalists do not have the theoretical resources to capture the *right kind of mistake* (genuine moral error, rather than error about the facts or error about one's feelings). This gives us strong reason to seek a normatively richer account of moral judgment than emotionism.

I argue that rationalists about moral judgment need not feel threatened by the importance of affect for moral judgment. I take it that the emotionist challenge does not merely say that emotions somehow accompany moral judgment, but that they *constitute* it. In order to make this constitution claim empirically plausible, one has to show that emotions are both *necessary* and *sufficient* for moral judgment. The following two chapters deal with the necessity and the sufficiency thesis, respectively.

In the fifth chapter, I review the empirical evidence for the claim that emotions are *necessary* for moral judgment. Research on psychopathy suggests that emotional impairments result in an inability to grasp crucial features of moral cognition, such as the moral/conventional distinction. I argue that the rationalist can wholeheartedly accept the necessity thesis. Although emotions are not perceptions of value, they play the same role for moral judgment as sensory perceptions play for judgments about the external world. Moral judgments without emotions are empty, emotions without moral reasoning are blind. This complements the idea that moral judgments are educated and rationally amenable moral intuitions because these intuitions are, at the core, emotionally charged quasi-perceptual seemings of what morality requires.

In the sixth chapter, I review the empirical evidence for the claim that emotions are *sufficient* for moral judgment. Emotion induction/manipulation experiments suggest that emotional changes alone can account for changes in people's moral verdicts. I argue that the rationalist can accept the sufficiency thesis as well. According to my reading of the thesis, the thesis that emotions are sufficient for moral judgment involves a causal claim; but there are conceptual constraints on whether a particular emotion causes a *genuinely moral* judgment or merely a reaction of emotional disapproval (like disgust or horror). An emotional reaction can only count as being responsible for a genuine moral judgment if it picks up on the morally relevant features of the situation *or* if the judging subject is willing to reconsider her judgment under improved conditions (and to either give it up or back it up with appropriate further grounding). Emotions only produce *genuine* moral judgments, rather than mere reactions of disgust or compassion, when they pick up on the morally relevant features of the situation. But when they do, they also confer sensitivity to moral reasons on the judgments they yield. Rationalism requires nothing more.

These two chapters are in many ways connected to the previous ones. Emotionally charged intuitions are a lot like moral perception: they provide subjects with an immediate sense of what is right and wrong. Just like ordinary perceptions, these emotional intuitions are trained to respond to particular morally relevant factors. And just like the deliverances of our senses, the validity of our moral intuitions can be called into question afterwards, thereby creating a need to provide adequate justification for their content—a need which is not there under ordinary circumstances.

My account is about how emotion, intuition, and reason interact. Joshua Greene, whose dual process model of moral cognition will be the focus of the third part, uses empirical evidence from neuroimaging, brain lesion studies, emotion manipulation experiments, and response time analysis to support the idea that there are two basic types of moral judgments: deontological and consequentialist, and that these two are produced by distinct cognitive systems—a quick, crude, emotionally charged system, and a slow, controlled, careful system. I will show that the empirical evidence there is for the claim that the distinction between the two cognitive systems can be mapped onto the distinction between these two types of moral judgments does not fare well under closer scrutiny. Along the way, I will also have the opportunity to revisit many of the themes of this book: the claims that emotion and reason exclude each other, that automatic judgments are somehow normatively dubious and consciously controlled judgments aren't, and that people's efforts at conscious reasoning are often hopelessly ineffective and sometimes thoroughly inaccurate.

6 What Is This Thing Called Morality?

The concept of morality is very hard to define. Therefore, I do not. The account I develop is supposed to be neutral with respect to theories of normative ethics. In principle, one could be a consequentialist or a Kantian, a virtue ethicist or a divine command theorist and accept most of my metaethical claims about the psychology of moral judgment. I do not know whether this is a virtue or a vice of my account, but besides the obvious advantages this fact has marketing-wise, there is a genuine philosophical rationale for the inclusive nature of the theory—it is dialectically more prudent. If I were to start with a consequentialist notion of morality that focuses on harm, my theory could be charged with a liberal Western bias. If

I were to start with a more deontologically flavored definition that focuses on rights and duties, others would be upset. And legitimately so, because metaethical theories ought to be about what moral judgments are, how they work, and whether they *can* be correct, rather than which moral judgments *are* correct. What I wish to say is that whatever morality turns out to require, here is how it works, and how it works is adequately described in rationalist terms.

I hardly ever say anything about what I mean by "morality" or the "moral" in "moral judgment," and I do not offer any explicit definitions of these concepts. I am aware that this is a serious flaw indeed, but I think it the lesser of two evils, and very well worth the price. If I were to define "morality," I would perhaps use the following definition: Morality is the set of the highest overriding principles governing human action. But with such a definition in hand, a vast proportion of research in moral psychology would become literally unintelligible, and appear not to be talking about morality (so understood) at all, but about an entirely different subject matter such as the behavioral effects of disgust, or the prosocial effects of empathy. Not having a prefabricated definition of morality is thus something like a necessary precondition for engaging with my topic at all.

Empirical moral psychologists constantly have to navigate between psychological inclusiveness—which requires conceptual lenience—and contemporary sophistication—which requires conceptual rigor. Nowadays, we distinguish between the rules of religious practice, etiquette, mere conventions, the values of various groups, cultures and subcultures, norms of theoretical and practical reason and, not least, genuinely *moral* norms. But what if fifteenth-century villagers did not make those distinctions? And what if, even today, many people remain unwilling (or unable) to see the same difference we see between the norms of their community, the obligations one has toward a deity, the rules of conduct that govern food, sex, death, and hygiene, loyalty to traditions and authorities, and the—allegedly *exclusively* moral—rules that tell us, basically, to let people do whatever they want as long as no one is harmed? Should we say that these people do not really understand what morality is about? Is it plausible to think previous generations did not make moral judgments? I do not think so. If we want to understand the psychological profile of moral cognition, we must sometimes bracket our own normative convictions.

Let me illustrate this a little further with an analogy to science and scientific beliefs. We are tempted to say that science is about the acquisition of reliable knowledge on the basis of observation, measurement, and experiment, all of those ideally couched in mathematical terms. But this practice was virtually absent in premodern times, and yet it would be ludicrous to suppose that there were no scientific endeavors. It seems clear that people were wrong about many things; that they were confused about how to explain what went on around them, or which methods for gathering and assessing evidence to use; that they often made ill-founded assumptions, told fictional stories, relied on hearsay or simply guessed how things had to work. This does not mean, however, that Aristotelian physics and alchemy are not proper parts of the history of science. The same holds for morality: *we* know, of course, that there is a meaningful distinction between religion, etiquette, and morality. But that people failed and fail to appreciate those distinctions, that they were confused and wrong about countless moral issues does not mean that they did not engage in the same practice as we do, or that they did not make any genuine moral judgments.

When doing moral psychology, one must always be wary of smuggling normative claims into seemingly descriptive ones. This is a problem to which many theories fall prey, especially when they try to take a decisive stand on whether moral judgment is based on emotion or reason. Andrew Sneddon (2011) makes the valuable observation that ultimately, the question which *sole* faculty moral judgments are really based on cannot be answered on empirical psychological grounds alone. Theories that try to identify one *core* capacity responsible for moral judgment are really making a moral point rather than a psychological one. He gives the example of an adolescent man who prefers moral judgments made from the gut rather than sober reflection; accordingly, the moral judgments he makes are intuitive to a significant degree. But the same man, once matured, might have the opposite preference, and value cool deliberation over the hothead views of his younger self. In the end, the question whether the core of our capacity to make moral judgments should be sought in our affective or our intellectual capacities boils down to the evaluative question, Which type of moral judgment seems preferable? This is yet another reason why I have decided to remain nonpartisan about the "true" nature of morality as possible.

I hope these last remarks will appease those unhappy with how I began this introduction. Some might object that becoming angry at somebody's table manners is not really a *moral* judgment, and I agree. But then again, only that which has no history can be defined, and the reality of moral judgment is more complex than our meager concepts would have us believe. What once was considered a moral issue might turn into a question of etiquette, and many things people once moralized about are nowadays seen as matters of personal taste. And how better to start a book about the rationality of moral judgment than with a story that reminds us of the immense progress we've made?

1 The Anti-Rationalist Challenge

Two ideas in recent empirically informed metaethics inspire skepticism about the rationality of moral judgment. One is about how *much* influence emotions have on our moral views and actions; the other is about how *little* influence reason has on them. This book is about why the tight relation between emotion and morality does not threaten the project of moral rationalism, and about how we ought to understand human moral reasoning, and its relation to moral judgment.

To appreciate the main thrust of the anti-rationalist challenge, let me start with some remarks about reasoning and decision making. Here is some preliminary evidence supporting the idea that we routinely overestimate the importance of reasoning for (moral) judgment:

(1) *Reasoning sometimes plays a sham role for judgment and decision making.* Humans have an astonishing ability to make up plausible sounding but wholly inaccurate stories to explain their beliefs and motivations to themselves and others, as well as to offer reasons for their judgments and behavior that could not possibly have played a role in their formation. This can happen, for example, when people are made to perform unintelligible actions through posthypnotic suggestion. Albert Moll, for instance, told one hypnotized subject to take a flowerpot from the window sill, wrap it in a piece of cloth, put it on a nearby sofa, and bow to it three times. The subject did as he was told. When asked why he did what he did, he replied that he wanted to keep the plant warm to prevent it from dying, and that he bowed to it because he "was pleased with [himself] for having such a bright idea" (Moll 1889, 153–154; see also Wegner 2002, 149). Making up demonstrably false or inaccurate stories and justifications like this is called *confabulation* (Hirstein 2005).

Michael Gazzaniga's studies with so-called split-brain patients (whose corpus callosum has been severed) show that there are other ways to dissociate the genuine causes of people's behavior from the justifications they offer after the fact. Patients who have undergone this type of callosal section can be selectively induced with stimuli that are processed only by one hemisphere. In one experiment, Gazzaniga and his colleagues flashed the word "walk" to a patient's left visual field—and, accordingly, his right cerebral hemisphere. The patient got up and walked away. Asked why he had done this, he replied that he wanted to get a Coke (Gazzaniga 1983). The verbally competent left half of his brain had made up a rationalization for his behavior that was entirely disconnected from the real origin of his intention.

Most cases of confabulation are less severe—but perhaps more interesting—than the examples taken from split-brain patients and hypnotized botanophiles. In a famous study by Robert Nisbett and Timothy Wilson (1977, 1978), subjects were presented with four pairs of stockings they could choose from. Unbeknownst to them, all samples were identical; but when asked why they had picked one rather than another, they mentioned the quality of the fabric or the superiority of the design of their preferred item. In fact, however, most subjects simply went for the pair that was on the right side of the display, but because subjects were unaware of this, they had to come up with something else.

(2) *Reasoning sometimes plays no role at all in judgment and decision-making.* Sometimes, people do not even bother to confabulate any justification for their beliefs. The sociologist Robert Bellah (1985) conducted a series of interviews about people's foundational moral values. When asked about the reasons for his endorsement of some general moral principles, one interviewee replied "Why is integrity important and lying bad? I don't know. It just is. It's just so basic. I don't want to be bothered with challenging that" (7). This explicit concession not to be able to provide any justification for one's beliefs seems to be the exception. Most of the time, and especially when they are interrogated by an authority figure such as a scientist, people feel the need to make some sense of their judgments. And when, as a matter of fact, no real sense is to be made of them, they invent some.

(3) *Reasoning sometimes interferes with judgment and decision making.* These are cases in which people's reasoning about their judgments and decisions clearly plays no causally effective role. In other cases, it does play

such a role, albeit a detrimental one. In one study, Timothy Wilson and Jonathan Schooler gave subjects several brands of strawberry jams and compared their favorites to the verdicts of a panel of experts. One group had to explain their choice; the other did not. But the judgments of those who were asked to analyze and list reasons for their preference conflicted more strongly with the opinion of the sensory experts than the judgments of the others. Making moral judgments and "forming preferences is akin to riding a bicycle; we can do it easily but cannot easily explain how" (1991, 182). And even if we have to while we are at it, it need not improve our performance. Later on in this book, I will discuss some experiments that study people's moral reasoning more specifically, rather than their judgment and decision making in general. The lesson, however, seems to stay the same: conscious reasoning of the kind that rationalists put such high hopes on does little work; emotions and unconscious influences carry the day.

On the other hand, it is undeniable that very often reasoning plays an indispensable role in moral judgment. Consider the following two examples:

(1) *Facts*. All particular moral judgments—that is, judgments about concrete moral issues rather than highly general and formal principles structuring those judgments—involve some factual component. In many cases, the values one endorses and the norms one accepts do not settle one's take on particular morally salient issues. Often, a lot of reasoning is required for someone to find out what one's values entail in a particular situation. Take the question whether rich people should be taxed more or less. How one answers this question will depend in many important ways on how highly one values property rights, social solidarity, and so forth. But it will also matter greatly how one assesses the facts, and how one's values translate to particular moral judgments in the light of those facts. Suppose someone is against higher tax burdens for the wealthy because she thinks that in the long run, these increases would undermine social solidarity. Or consider someone who supports higher taxes for rich people, because she thinks this would help the universal enforcement of property rights for all. Elaborate conscious reasoning is often necessary for people to find out what their values imply.

(2) *Principles*. Although some are denying that moral principles play any useful role for moral judgment at all (Dancy 2004; see also McKeever

and Ridge 2006), it is clear that, as a matter of fact, many people do employ such principles either implicitly or explicitly in their moral cognition. But principles, whether moral or not, need to be applied to particular cases, and they certainly do not apply themselves. This is where reasoning comes into play. Whenever there is a moral principle such as *Do not lie* or *Help others in need*, we need our capacity to reason to figure out whether a given case is an instance of the type of action the principle assigns a certain evaluative or deontic status to.

The bottom line is that the relationship between reason and moral judgment is a complex one, and that no simple account of it seems to be forthcoming. The theory I aim to develop in this book is supposed to do justice to this complexity.

1 The Effectiveness of Moral Reasoning

Introduction

Skepticism about the impact of reasoning on our thoughts and actions is not new. Here is Nietzsche's take on the issue:

> There are systems of morals which are meant to justify their author in the eyes of other people; other systems of morals are meant to tranquilize him, and make him self-satisfied; with other systems he wants to crucify and humble himself, with others he wishes to take revenge, with others to conceal himself, with others to glorify himself and give superiority and distinction,—this system of morals helps its author to forget, that system makes him, or something of him, forgotten, many a moralist would like to exercise power and creative arbitrariness over mankind. ... In short, systems of morals are only a sign-language of the affects. ([1886] 1966, 187, translation slightly amended)

Many moral psychologists working today would agree with this.

The theoretical framework that best exemplifies the anti-rationalist challenge is Jonathan Haidt's social intuitionist model (henceforth: SI model) of moral judgment. This chapter discusses the main claim regarding the nature of moral reasoning that rationalism about the psychology of moral reasoning is committed to (1) and the psychological paradigm that best exemplifies that claim (2). Current anti-rationalist theories, like the SI model, try to undermine the central tenets of this view (3) and raise the worry that moral judgments might not be based on reasons at all (4). Anti-rationalists develop an interpretation of the relation between moral judgment and moral reasoning that does not rely on a strong notion of practical reason, arguing that moral verdicts are based on emotionally triggered intuitions. Moral reasoning, they hold, plays no formative role in people's judgments whatsoever, but typically consists in mere confabulation (5). I discuss the details of this picture of moral reasoning and some of the most prominent objections to it (6). There is empirical evidence that the phenomenon of confabulation can be observed in how people justify their beliefs and actions, as well as in how they respond to the justifications of others (7), and that principle-based moral reasoning is not immune to *post hoc* rationalization, either (8). In the last three sections, I will show that the anti-rationalist challenge the SI model makes crucially rests on two requirements a process of moral judgment formation must meet in order to count as *rational*, and explain why these requirements can be rejected. This will allow me to develop an account of the influence of reasoning on moral judgment that can do without those requirements in the next chapter.

1 The Effectiveness Thesis

Rationalism about the psychology of moral judgment makes empirical claims. In particular, it holds that moral judgments are based on (good) reasons, and that what good reasons one has makes a psychologically real difference with respect to the moral judgments one makes. What are good reasons, as opposed to other kinds of reasons? We can distinguish the reasons that bring a person to make a certain moral judgment from the reasons that (would) ultimately justify it. This is commonly referred to as the distinction between a person's *motivating* and *normative* reasons. But what is the connection between these kinds of reasons when it comes to the normative reasons one has for a moral judgment, and the motivating reasons that actually cause the person to make it? At first sight, it seems that the answer to this question is fairly straightforward: we have motivating and normative reasons, and sometimes our judgments are based on the first, sometimes on the second kind. They are, however, potentially in conflict. If we think about the fact that our motivating and our normative reasons do not necessarily have to be the same, and that, provided a judgment has been made, there always has to be a motivating reason involved that is, as it were, responsible for what happened, we can ask whether our normative reasons ever are our motivating reasons, or whether we ever are motivated to do or believe something by what also justifies our actions or beliefs.

A negative answer to this question amounts to the claim that, when we decide what to do or to think, the normative reasons we have for our practical judgments are not part of the explanation for why we make these judgments. This would entail that the "deliberative perspective," as Michael Smith calls it (1994, 131)—the perspective on which normative reasons a person has—is not at the same time a causal perspective—a perspective that explains why that person does what she does and thinks what she thinks. Smith holds that to deny that the normative reasons we have for our practical judgments are part of the explanation for why we make these judgments would be "patently absurd":

> In order to see that the deliberative perspective is indeed a perspective on explanation, it suffices to note that to say otherwise is tantamount to supposing that the connection between what we decide to do, on the basis of rational deliberation, and what we do, is altogether contingent and fortuitous. ... When we deliberate, and decide what we have a rational justification for doing, that very fact sometimes makes a difference to what we do. (132)

The rationalist wants to maintain that when we decide what to do or to think, our normative reasons sometimes make a difference (cf. Smith 1994: 132; see also Korsgaard 1996; Scanlon 1998; Velleman 2000). We can call this the effectiveness thesis:

Effectiveness Thesis (ET)

The justifying (moral) reasons subjects have for their (moral) judgments figure in true causal explanations for why they hold these judgments.

The thesis states that episodes of practical and, more importantly, moral reasoning—the practice of giving justifying reasons for moral judgments—are causally effective.[1] This is the most central claim rationalist accounts of the psychology of moral judgment make.

One question that immediately comes to mind is this: how often does this actually happen? Surely, rationalists would want to allow for the possibility of people's reasoning being ineffective; at least sometimes, rationalists must be willing to grant, people rationalize rather than reason properly. It is important to bear in mind that discussions about the psychology of moral reasoning are not about the simple quantitative issue of *how often* normative reasons become causally effective or not. Rather, it is about whether they do in typical core cases: whether it happens *often enough* for it to make sense to hold people to the demand that their judgments be based on genuine moral reasoning.

The present chapter will mostly be concerned with the evidence for the *in*effectiveness of moral reasoning. This might come as a surprise to some, but there is a rationale for this, which is to firmly establish that rationalizations of moral judgment and other forms of decision making are a genuine phenomenon that cannot be explained away either empirically (the evidence is flawed) or normatively (the evidence is normatively irrelevant). After this, the stage will be set for me to argue that the available evidence gives us no reason to think that typical core cases of moral reasoning are confabulatory. This, I will argue, is everything the rationalist could—and should—ever ask for.

2 The Conscious-Reasoning Paradigm

The psychological account of moral judgment and reasoning that best exemplifies ET is Kohlberg's developmental model of moral cognition. Moral psychology in the Kohlbergian tradition focuses on the way people justify their judgments in response to moral dilemmas and tries to discover the developmental patterns behind the reasoning that is performed by subjects. For instance, a justification for a normative judgment that essentially refers to possible punishment ("A-ing is wrong because I will be grounded") is considered to display a lower stage of *moral maturity* than a justification that has an essential reference to peer-group loyalty ("A-ing is wrong because I don't want to let my buddies down"). And, those kinds of justifications, again, are considered to display a lower stage than *moral reasoning* on the basis of abstract and universal norms and principles like human dignity, the social contract, or the categorical imperative (Kohlberg 1969; see also Piaget 1965). What is crucial for this psychological paradigm is that it studies subjects' performances with an attitude of charity: it presupposes that, at least in principle, human beings have the capacity to arrive at correct moral judgments and that there can be progress in moral consciousness from the primordial, egocentric moral thinking of a young child, to sophisticated and well-founded adult moral reflection. But no matter how primitive the reasoning abilities subjects exhibit, the assumption is never questioned that it is people's explicit reasoning that is also causally responsible for their judgments. (This holds, with some qualifications, even for Kohlberg's most prominent critics; cf. Gilligan 1982 and Turiel 1983.) It is precisely this assumption that is questioned by recent empirical moral psychology.

3 Running Out of Reasons

It is a well-established fact in cognitive psychology and neuroscience that under certain (usually pathological) conditions people can be staggeringly confident about states of affairs which obviously do not obtain. Patients with Korsakoff's syndrome report in great detail about trips they have made and conferences they have attended, when in fact they have not left the hospital for months. People with Capgras syndrome confidently claim that one or more of the people close to them have been replaced with impostors, and make up explanations for who might have done such a thing to them, and why. Characteristically, these patients do not withdraw their confidence when their mistakes are pointed out to them. Instead, they start to confabulate, an activity that is motivated by "a sort of pathological certainty about ill-grounded thoughts and utterances" (Hirstein 2005, 4). Speaking from the perspective of cognitive neuroscience, the phenomenon of confabulation suggests that there are two separate mental processes: one *creating* explanations, and another *checking* whether they are true.

What does this have to do with moral reasoning? The SI model rejects the conscious-reasoning paradigm of moral deliberation as utterly unrealistic and maintains that conscious reasoning usually doesn't play a role in the formation of moral judgment at all. Rather, social intuitionists argue, arriving at a moral evaluation is a largely subconscious process of quick and effortless emotional information processing, and our moral judgments are based on intuitive "gut reactions" about the core themes of morality. The obvious fact that people do engage in explicit and conscious moral reasoning can be squared with that claim by showing that episodes of conscious reasoning are *post hoc*: they are biased, after-the-fact rationalizations of intuitively formed beliefs that serve to justify the moral convictions people already have, independent of the reasons that support them logically (Nisbett and Wilson 1977, 1978). Conscious moral reasoning—the weighing of the normative reasons in favor of a judgment—cannot be the source of the motivating reasons—the mental processes that actually lead people to adopt a certain judgment—if it comes after the fact. This can be called the *post hoc* thesis:

Post Hoc Thesis

When people engage in moral reasoning, they do so after a moral judgment is reached.

The most striking support for this claim comes from the phenomenon of "moral dumbfounding." For people's most central moral convictions about, say, incest, cannibalism, abortion, or torture, it can be shown that subjects stick to their judgments about particular morally significant situations even if their reasons do not to apply to the situations at hand. One handy way to study the impact of moral reasoning on moral judgment is to confront people with fictitious yet more or less realistic scenarios or thought experiments, and ask them for a moral assessment of the described cases. One of those scenarios might read something like this:

Incest I

Julie and Mark are brother and sister. They are traveling together in France on summer vacation from college. One night they are staying alone in a cabin near the beach. They decide that it would be interesting and fun if they tried making love. ... What do you think about that, was it OK for them to make love? (Haidt 2001, 814)

Most people are inclined to give a clear and definitive answer to that question: "No." Let's assume, for the sake of the argument, that this answer is correct. And let's also assume that people are ready to provide all kinds of reasons for their moral judgment that it was not OK for the siblings to make love; for example, they might argue that sexual acts between brother and sister can threaten the stability and the emotional infrastructure of the special relationship between members of a family; or that sex between near relatives carries the risk of severely handicapped children; or that incest jeopardizes the most basic family values; or maybe even that God just doesn't approve of it.

Jonathan Haidt's social intuitionism is based on the ingenious idea to confront test subjects with "tweaked" moral dilemmas that elicit an immediate emotional response and subconsciously trigger a moral judgment; however, almost all possible justifications—whose developmental level the conscious-reasoning paradigm and Lawrence Kohlberg are interested in—have been removed from the story up front. The prediction that best coheres with broadly rationalist explanations of the practice of moral judgment is that once this is pointed out to test subjects, they will withdraw their judgment or back it up with appropriate further grounding.

But this is not what actually happens. This is the complete version of the above story, the one that *really* was presented to subjects in Haidt's research about the fundamental shortcomings of human reasoning about moral issues:

Incest II

Julie and Mark are brother and sister. They are traveling together in France on summer vacation from college. One night they are staying alone in a cabin near the beach. They decide that it would be interesting and fun if they tried making love. At very least it would be a new experience for each of them. Julie was already taking birth control pills, but Mark uses a condom too, just to be safe. They both enjoy making love, but they decide not to do it again. They keep that night as a special secret, which makes them feel even closer to each other. What do you think about that, was it OK for them to make love? (Haidt 2001, 814)

Surprisingly, subjects make the same judgment when presented with this vignette: no, it was not "OK" for them to make love. When asked for a justification for their judgment, they respond in the familiar way. Even though none of their justifying reasons apply to the scenario, they stick to their judgment and reasoning, and this calls for an explanation.

Recent moral psychology claims that moral judgments—or at least an important subclass of moral judgments—are based on emotion rather than reason. This view, however, owes us an explanation of why such a superfluous thing as moral reasoning occurs in the first place. This is where the concept of confabulation comes into play. Many moral psychologists nowadays claim that explicit moral reasoning is like confabulation: it is like making up excuses for beliefs that are held not because they are factually accurate, but in order to avoid emotional disturbance in one's values. In short, they deny that ET holds.

The picture offered by the SI model suggests that moral reasoning is some kind of confabulation. Rather than giving an empirically accurate description of how they arrived at their judgment, moral reasoning works like telling a story which does not necessarily have to be connected to what was really going on. Our emotionally triggered intuitive reactions—whether they are based on negative feelings like disgust or positive feelings like empathy—are in fact sufficient to explain our practice of moral judgment, rendering explicit reasoning utterly superfluous (Wheatley and Haidt 2005; Schnall et al. 2008).

Let me briefly reframe the challenge the SI model poses in terms of a two-systems theory of moral judgment, the basic outlines of which were introduced in the introduction. Recall that dual-process theories hold that mental processes fall under two major categories: System I processes are evolutionarily old, work automatically, quickly, effortlessly, and unconsciously, and are often emotionally valenced. System II processes, on the other hand, are emotionally cold, slowly undertaken, cognitively demanding, and deliberately controlled processes that have not been developed until very recently and are generally thought to be uniquely human (like the capacity for long-term planning and counterfactual reasoning). The claim about moral reasoning marshaled by the SI model, then, is that while it appears to people that their moral judgments are based on System II processes, the empirical evidence tells us that the opposite is the case. What is going on *in the mind* is largely dependent on rapid, unconscious System I processes, with System II not genuinely affecting, but at the most registering and reporting, what is going on behind the locked doors of the unconscious. Indeed, according to the SI model, this is what creates the illusion of conscious reasoning in the first place, because since it is only the result of the process that becomes consciously accessible, it subjectively appears as if these very episodes of conscious thinking were in fact the decisive element of the process.

On the basis of the empirical evidence about how ordinary people react once their reasons are debunked, Haidt concludes that a subject's justifying reasons bear no effective connection to his moral beliefs. But does the model allow for exceptions to this rule? Remember that the effectiveness thesis claims that our justifying moral reasons make a causal difference (in core cases of moral judgment). Can the model account for these—allegedly rare—cases where it does happen that genuine normative reasoning has a causal influence on people's moral judgments? Does the SI model pose a *global* challenge to the effectiveness thesis, or is there a place left in the model for genuine normative reasoning?

4 The *No Reasons* Interpretation

Perhaps the intuitively most appealing view about the connection between moral judgment and moral reasoning is this: upon encountering a situation that calls for a moral judgment, people consciously weigh up all the available reasons and settle for the judgment that is favored in their light. The SI model rejects this picture as utterly unrealistic and maintains that conscious reasoning usually doesn't play a role in the formation of moral judgment. According to this framework, the effectiveness thesis appears to be overly optimistic. Using Hegel's famous metaphor, we can say that according to the SI model, moral reasoning is like Minerva's owl: it starts its flight at dusk, when the job is already done.

How are we to interpret this remarkable finding? It seems that if people do not withdraw their judgment when it is pointed out to them that their reasons are invalid, we have reason to assume that even in the cases where this cannot be done, (perhaps because the reasons that justify their judgments are in fact valid) reasons do not play an effective role in the way people arrive at their judgments. Jesse Prinz writes:

> The problem ... is that people usually don't revise their moral assessment when their reasons are debunked. ... This suggests that the reasons they offer did not play a very central role in the formation or maintenance of their moral judgments. (2007, 31)

And further:

> A ... possibility is that subjects have *no reasons* for their moral judgments. They simply have a gut reaction that consensual incest and laboratory cannibalism are wrong, and a few post hoc rationalizations, which play no important role in driving those reactions. (31)

Prinz holds that the norms and values that people refer to with their gut reactions are basic in a *justificatory* sense. They are placed "outside the reason-giving game" and morally dumbfounded people have simply hit "rock bottom." Call this the *no reasons* interpretation.

Let me mention two main problems with this interpretation. First, it may well be that there are moral beliefs that are properly basic in a justificatory sense. This claim would entail that people are entitled to hold certain beliefs about what is morally right or wrong without being obliged to cite any explicit evidence to support them. The fact that many of subjects' judgments are based on no reasons at all would not pose a problem on this view, because moral values rest on an emotional foundation for which no rational justification is to be expected anyway. But the justificatory basicness of moral intuitions only holds in the *absence* of legitimate challenges to their judgments, that is, as long as there is no positive reason for people to think that they are mistaken in their belief. *Incest II*, however, is a stone pit of possible challenges, and so the "no reasons" interpretation doesn't work here.

Second, the "no reasons" interpretation conceives of a belief's justificatory basicness in a merely descriptive way. To really count as a foundation for other beliefs—one that is strong enough to carry them, as it were—it is not enough to have a set of beliefs that, as a matter of fact, are *treated* as standing at the end of a justificatory chain. There is no guarantee that these basic beliefs really do convey the rational force necessary for them to serve as a solid foundation for other moral judgments. Subjects' tendency to treat many of their foundational moral intuitions as not standing in need of justification could just as easily be explained in terms of nonrational biases that would motivate them to accept many moral views in good faith even if they were not true.

One of my main aims will be to offer a response to the no reasons interpretation. I will suggest that reasons and reasoning do play a formative role for people's moral judgment, namely through the education of their moral intuitions, over the course of which their reasoning becomes habitualized over time. Before I explain the details of this account, let me take a closer look at the structure of the SI model, and the precise target of its anti-rationalist challenge.

5 Moral Reasoning as Confabulation

In Haidt's experiments, people run out of reasons fairly quickly. Once this has been pointed out to them, most people tend to claim a kind of self-evidential status for their basic moral intuitions. The explanation that is favored by recent empirical moral psychology is that people do not arrive at their judgments on the basis of explicit moral reasoning in the first place. If they did, why don't they alter their judgment in cases where their reasons don't apply? People's chain of reasoning ends with what Haidt calls moral "dumbfounding." Confused, subjects smile as if they feel caught out, and claim that a certain action is "just wrong" and that, somehow, they know it is.

The picture offered by the SI model suggests that moral reasoning is confabulatory. Rather than giving an empirically accurate description of how one arrived at one's judgment, moral reasoning is like coming up with a story that does not necessarily have to be connected to what really obtains in the case at hand. Call this the confabulation thesis:

Confabulation Thesis

Moral reasoning is a matter of causally ineffective confabulation.

Our emotionally triggered intuitive reactions—whether they are based on negative feelings like disgust or positive feelings like empathy—are in fact sufficient to explain our practice of moral judgment, rendering explicit reasoning utterly superfluous (Wheatley and Haidt 2005; Schnall et al. 2008). Let me emphasize again that according to their most charitable interpretation, neither the effectiveness nor the confabulation thesis makes straightforward quantitative claims. They do not state that moral reasoning is *often* causally effective, or that it is *mostly* a matter of mere confabulation. Rather, they say that in paradigmatic core instances, this is the case. I will return to this point below. For now, let it suffice to say that it is notoriously difficult to pin down exactly what rationalist and anti-rationalist models of moral judgment are committed to. As I see it, the dialectically most fruitful way to deal with this issue is to attribute claims that are not obviously false or trivially true to both positions, regardless of whether this attributed position can be found in the explicit statements of the respective views. An obviously false rendition of rationalism, for instance, would be that all our moral judgments are produced by conscious reasoning; a trivial interpretation of rationalism would be that this can sometimes happen. An obviously false reading of the SI model would be that neither proximal nor distal reasoning can ever have any impact whatsoever on people's moral judgment; a trivial interpretation of the model would have it that reasoning is sometimes causally ineffective.

It is useful to think of the anti-rationalist program as a "bypassing" challenge. The core idea behind such challenges is that in some cases, the inferential sequence subjects cite as grounds for a given belief or action did not actually play a causal role in why they hold that belief, or performed that particular action (see figure 1.1). In this case, the moral judgment that *incest is wrong*—either in this particular case or in general—may well be justified by the reasons specified in the set of premises. But the underlying causal sequence may tell rather a different story.

The *post hoc* thesis and the confabulation thesis need to be carefully distinguished. The former is compatible with the fact that people simply make explicit the reasons that *really did* play a formative role in how they arrived at their judgment, albeit automatically and unconsciously. Only the latter actually denies ET.

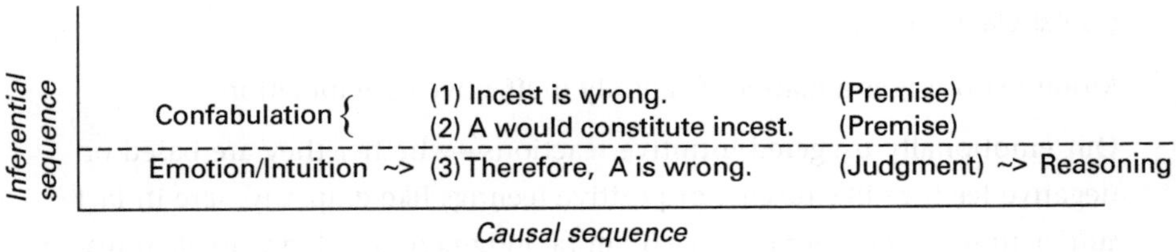

Figure 1.1
Anti-rationalism as a "bypassing" challenge.

Empirical evidence about how people actually reason suggests that they come up with reasons to justify their moral judgment after the fact. From this, intuitionist models generalize to the claim that typically, episodes of moral reasoning are a matter of mere confabulation, which entails that ET is not true. In fact, however, the evidence for the *post hoc* nature of moral reasoning is compatible with ET because the confabulation thesis *only* follows from the *post hoc* thesis *in combination with two problematic assumptions we are given no independent reason to accept.* Before I turn to those assumptions, let me spell out the intuitionist challenge and Haidt's responses to anticipated rationalist objections to his model in a little more detail.

6 Moral Reasoning from an Intuitionist Perspective

The SI model distinguishes between six "links" or "processes" by which a subject can arrive at a moral judgment (Haidt 2001, 818). Moral reasoning is defined as a "conscious mental activity that consists of transforming given information about people in order to reach a moral judgment" (818). This process is "intentional, effortful, and controllable" and the reasoner "is aware that it is going on" (818). Haidt stresses that the model does not entail that no such thing as moral reasoning exists: four of the six links, he insists, are moral reasoning links. Critics of his model, however, hold that the model cannot account for genuine *moral* reasoning, but merely for a correction of the facts moral beliefs are based on and the nonrational triggering of other intuitions. What role can really be played by moral reasoning, according to the SI model?

(i) Given that he understands the question of whether there is any such thing as genuine moral reasoning in terms of whether moral judgments are *caused* by conscious moral reasoning (an idea I will criticize below), this is what he is looking for: "The core of the model gives moral reasoning a causal role in moral judgment but only when reasoning runs through *other people*" (819). But this looks like mere persuasion rather than moral reasoning, which is something that won't satisfy the rationalist. Moral reasoning, he argues, cannot be considered as something that is merely based on social pressure, but rather by the uncoercive coercion of the better argument. Does the social intuitionist model leave room for that?

(ii) Private reasoning can also play a role for moral judgment: "In the course of thinking about a situation a person may spontaneously activate a new intuition that contradicts the initial intuitive judgment" (819). Again, this is not the kind of reasoning one was looking for. What Haidt has in mind seems to be a "hydraulic," competitive picture of how the process of weighing between competing intuitions works. Intuitions show up in the mind; perhaps elicited by a person's closer examination of a case and his own thoughts about it, they carry a particular strength, and they combat for the leading role in a person's set of intuitions. But the description of this as a process of "spontaneous activation" and a "triggering" of "new intuitions" runs counter to the implications of a strong notion of practical reason. As Saltzstein and Kasachkoff put it: "Haidt concedes that people do change their minds about moral matters but insists that the change is not a reasoned one; rather, it is a nonrational response to the triggering of new intuitions" (2004, 278).

(iii) To block the worry that the SI model leaves room not for genuine reasoning, but merely for the nonrational triggering of new intuitions, Haidt has pointed to the fifth link in his model, the "reasoned judgment link": "People may at times reason their way to a judgment by sheer force of logic, overriding their initial intuition" (Haidt 2001, 819). Haidt claims the activation of this link is rare and most likely to occur where the initial intuition is weak, and the capacity to process relevant information high. But we can, the SI model admits, at times override our initial gut reaction toward, say, cannibalism or homosexuality in case we become aware of the fact that the justifications we are emotionally inclined to put forward are factually inaccurate. This is the most important concession Haidt makes to defenders of ET.

But to most rationalists about the psychology of moral reasoning, this is still unsatisfactory. There just seems to be no space in the web of our intuitions for genuine *practical* reason—which goes beyond the ability to take into consideration new factual information—to become effective, and for genuine *moral* justification to come into play (Saltzstein and Kasachkoff 2004; Musschenga 2008; Clarke 2008). A merely quantitative attenuation of the confabulation thesis does not solve the problem that the SI model has a hard time dealing with cases of genuinely *moral* reasoning.

Let me give an example. It is questionable, for instance, whether it is legitimate to refer to *nature* to justify your moral disapproval in the first place. One could make the point that, in the consensual incest case, liberal rights should take priority over so-called family values, or question the eugenic logic behind the argument that it is the possibility of handicapped children that prohibits incest. All these different types of reasoning go beyond the mere triggering of new intuitions, because (a) they are apt to question the very legitimacy of the intuitions we happen to have, and (b) they are not reducible to the acknowledgment of mere *factual* information. They bring new, genuinely moral considerations to bear on the issue at hand. The SI model is unable to account for moral reasoning of this type. Indeed, one need not mistake Haidt's "emotional dog" for a blind instinct "possum" (Haidt 2004) to see that his model is committed to an outright denial of the effectiveness thesis. It is worth mentioning that this is one major source of disagreement between the dual process and the SI model, as Paxton and Greene are happy to concede that consequentialist moral reasoning is not merely morally relevant, but in many cases also causally effective, especially when epistemic conditions (sufficient response time and information-processing capacities) are favorable: "If the SIM is correct, then attempting to engage others through their ability to reason, aiming messages at the 'head' rather than the 'heart,' is an exercise in futility. Indeed, depending on what one means by 'reasoning,' one might say that the SIM does not really allow for 'reasoned persuasion' at all. According to the SIM, we can say things in hopes of modifying one another's moral intuitions, but it is impossible to convince someone to put her moral intuitions aside and, instead, reach an alternative, counter-intuitive conclusion based on reasoning" (Paxton and Greene 2010, 514).

So far, I have articulated the structure of the anti-rationalist challenge and discussed some of the evidence for it. I will discuss even more of it in the two sections to come. Note that my aim in this chapter is not to dispute the empirical credentials of the SI model. Rather, it is to set the stage for my constructive response to the model, which can be found in the second and third chapter of this book. Therefore, I will review the evidence for moral reasoning straightforwardly, and only start criticizing the conceptual assumptions it has to be combined with for the SI models anti-rationalist conclusions to follow.

We have seen that *proximal* moral reasoning—that is, reasoning that is undertaken directly before and in order to arrive at a moral judgment—often plays little or no role at all in subjects' responses. My point will be that while this might be true, it actually poses no threat to rationalist models of moral judgment. We ought to be looking for the effectiveness of moral reasoning in *distal* reasoning, that is, reasoning that helped educate people's moral intuitions in the first place, as well as postjudgmental reasoning in response to challenges from within a social practice of moral reasoning. I will explain the role of such distal moral reasoning in more detail in chapters 2 and 3. Now, I shall briefly present some more findings on the phenomenon of confabulation in practical and moral reasoning.

7 Placebic Reasons

Research on the confabulation of moral reasons focuses on cases where people *give* ineffective reasons for their judgments. But to a large extent, reasoning is a social practice: it consists in giving reasons for your own judgments as well as *asking* for and receiving other people's reasons. The game of giving and asking for reasons is one in which people hold *each other* accountable for their judgments, and this includes the *production* just as much as the *reception* of reasons.

There is evidence suggesting that the reception of reasons is prone to something analogous to confabulation. It can be shown that under certain conditions, the reasons that one person is given by another person can influence the first person's behavior, even though the received reasons contain no significant information whatsoever for the receiving agent. To test the hypothesis whether this kind of pseudo-reasoning can actually have significant effects on people who are given pseudo-reasons, Ellen Langer and her colleagues (1978) conducted the following experiment.

They placed a confederate behind a table right next to a university's copying machine; test subjects were the people who just happened to use the machine the day the experiment was made. The idea of the experiment was to ask the people who were waiting in line for permission to use the copy machine and to either just ask them whether they were willing to do so ("May I use the copy machine?," without giving them any reasons), or to ask them and give them what Langer and her colleagues call "placebic" reasons ("[...], because I have to make copies?"), or to ask them and give sufficiently valid reasons ("[...], because I am in a hurry?"). A further variable was how big the favor was that the confederate asked for. Randomly assigned conditions differed between five and twenty copies that had to be made. In the big favor-condition, that is, when a greater loss of time was at stake, the redundant reason did not work all that well. In the small favor-condition, however, the placebic reason was just as effective as the valid reason, even though it conveyed no practically salient information whatsoever for the reason-recipient.

This suggests that behavior which appears to be guided by reasons can in fact work "mindlessly," on a thoroughly automatic level. The information that is processed on this level does not in any interesting way have to be connected to what is happening on the higher level of consciously processed information, that is, on the level where people's justifying reasons can be located.

One can easily see that this is a case that is analogous to confabulatory reason-*giving*. There is an episode of explicit reasoning, followed by a corresponding behavior that appears to be justified by the reasoning. But as a matter of fact, this cannot be what is going on: the received reasons *as such* cannot possibly have played a formative role in bringing about the behavior, since they did not convey any information that could have done that job. The reasons did not bring about the behavior in virtue of their content, by telling the reason-receiving person why it makes sense to do as the given reasons recommend. To justify a demand to use a copy machine first by saying that one wants to make copies is obviously redundant, and hence cannot exert any kind of rational force upon the addressee of the demand. What appears to be genuine—and, most importantly, causally effective—reasoning turns out to be a mindless social charade that may influence people's actions on a subconscious level, but completely bypasses their rational control and deliberation.

8 Moral Principles: Universal Moral Grammar or Confabulation?

Does the confabulation thesis apply to reasoning on the basis of general moral principles as well? Are people's normative reasons, to the extent that they partly employ moral principles, psychologically effective? Or is which principles one endorses a matter of unconscious influences and after-the-fact rationalizations as well?

There is evidence that general principles do in fact play a role when people make moral judgments (Cushman et al. 2006). In the trapdoor variation of the classic footbridge dilemma, most people think it impermissible to drop a large stranger onto the tracks in order to save five innocent people from a runaway trolley, even though—contrary to Joshua Greene's original explanation (Greene et al. 2001, 2004)—no emotionally troubling harm is involved because the person does not actually have to be *pushed* off the bridge (for a response to this problem, see Greene et al. 2009). This suggests that it is not merely the emotional impact of the fact that pushing the man is an act of directly hurting another human being with one's own bare hands, but because one is using him as a means to a certain end rather than a foreseen but unintended side effect. People generally find the former to be morally impermissible, and this is explained by a moral principle: the principle of double effect. This suggests that moral rules and emotional reactions are equally important for the practice of moral judgment (Nichols and Mallon 2006; Leslie et al. 2006). What the evidence also shows is that these principles often remain inaccessible to people's consciousness: subjects hardly ever explicitly refer to the doctrine of double effect to justify their judgments (Hauser et al. 2007). There can be a dissociation between *operative* and *expressed* moral principles.

However, it remains questionable whether people actually *follow* these principles in the sense that they have a stable repertoire of rules, as universal moral grammar theory (Mikhail 2007) suggests, which they unconsciously use to arrive at their judgments. There is evidence that the structural descriptions of cases on which, according to moral grammar theory, moral assessments are based sometimes rest on a prior moral assessment themselves. Most famously, Joshua Knobe (2003) showed that ascriptions of intentionality sometimes track normative judgments (bad side effects of actions are more likely to be considered intentional than good side effects), instead of the other way round.

There is further evidence that people are flexible when it comes to the question of which principle(s)—the consequentialist principle to maximize the overall good or the deontological principle that it is always impermissible to use another human being as a mere means to an end, even if it maximizes the overall welfare—to apply to a given moral problem. David Pizarro, Eric Luis Uhlmann, and others (Uhlmann et al. 2009) have shown that depending on the particular emotionally salient features of the case test subjects are confronted with, they are prepared to adjust their use of moral principles according to the emotionally triggered intuition they have. When pressed to offer a justification for their judgment, they start to confabulate why a particular principle applies to the case at hand. They do use moral principles in their reasoning, although not in the sense that they have a universal set of rules they adhere to, but by *choosing* the principle *post hoc* that best justifies the judgment they want to end up with.

A similar effect can be found in nonmoral—but obviously morally very relevant—cases such as a hiring decision, where people are happy to adjust their criteria for suitability for a position depending on implicit biases in favor of certain groups (Uhlmann and Cohen 2005). This suggests that it is often not the principle that determines the judgment, but the judgment that determines which principle is deemed appropriate. Again, the justificatory relation that certain moral principles might bear to the judgments that are being made does not, as the effectiveness thesis has it, play a causally effective role in the formation of people's judgments. Rather, as input into the actual process of deliberation, the principles are utterly superfluous, and people do not hold certain beliefs because they are justified in the light of the principles they endorse, but search for justifying principles because they hold certain moral beliefs.

Pizarro and his colleagues presented their test subjects with modified versions of the original footbridge-dilemma, which were manipulated in a way that contained tacit information about the race of the people who had to be killed or prevented from being killed:

> Half of the participants received a version of the scenario where the agent could choose to sacrifice an individual named "Tyrone Payton" to save 100 members of the New York Philharmonic, and the other half received a version where the agent could choose to sacrifice "Chip Ellsworth III" to save 100 members of the Harlem Jazz Orchestra. (Uhlmann et al. 2009, 482)

Before running the experiment, test subjects' political orientation (from *very liberal* to *very conservative*) was tested on the basis of self reports. Also, people were asked whether they thought that a person's race or nationality mattered to his or her moral standing (which, with respect to both features, 87 percent denied).

As it turned out, however, matters of race and nationality *did* have an effect on subjects' moral judgments. Generally speaking, people were much more likely to endorse consequentialist or deontological judgments and justifications if these judgments and justifying principles fit their prior political convictions. Liberals, for example, were more likely to sacrifice "Chip" and less likely to endorse sacrificing "Tyrone." Pizarro and his colleagues explain this differential application of moral principles with the strong antipathy against racial discrimination liberals share. Their judgment that it is permissible to sacrifice Chip for the sake of saving the members of Harlem Jazz Orchestra can thus be seen as an overcompensation in the name of racial fairness. Complementarily, conservatives were more likely to judge it permissible to kill innocent civilians in a military attack if the collateral damage affected Iraqis, but not so if it affected Americans. The most natural explanation for this seems to be that in this case, preferring one moral principle over another is due to the strong sense of community and in-group loyalty typically shared by conservatives.

At any rate, the relation between moral judgments and abstract moral principles is not as an overly simplistic rationalist picture of moral reasoning—the picture that moral judgments are made on the basis of a *mechanical* application of principles—suggests. The use of moral principles is often motivated by the not consciously accessible force of emotional reactions toward the particular features of a situation, although these responses might (and often do) pick up on features that—even according to the standards people themselves hold to be adequate—do not necessarily speak to the moral quality of the judged actions and events.

9 Moral Choice Blindness

An even deeper problem for folk reasoning about morally salient cases is that sometimes, subjects are unlikely to have any firm moral beliefs about many issues at all. Unfortunately, participants rarely admit (or perhaps even notice) this fact, and so they tend to make something up. This is exacerbated by the subtle social pressure of taking part in an experiment where people feel that they have to give at least *some* response, even if they do not have a genuine, worked out opinion on the topic at hand.

In a particularly intriguing study, Hall and his colleagues (2012) could show that people's willingness to simply invent responses, to readily change those responses, and to come up with confabulatory justifications for them can be quite dramatic. In this experiment, they used a sleight of hand design to demonstrate the frailty of people's moral intuitions. Participants were given a clipboard with a series of statements with either *general* moral content ("Even if an action might harm the innocent, it can still be morally permissible to perform it") or *specific* moral content ("The violence Israel used in the conflict with Hamas is morally defensible despite the civilian casualties suffered by the Palestinians"). They then had to indicate their agreement with those statements on a one-to-nine scale. After completing the first page of the survey, they had to flip it back to get to the second series of propositions.

Unbeknownst to them, a small patch had been glued to the back of the clipboard on which the content of some of the statements had been reversed (e.g., from permissible to impermissible). This patch was automatically picked up by the first page, so when they flipped it back again, the agreement they had indicated for the original statements had turned into their opposite. Not only did 69 percent of people not notice this change in content of at least one of their statements (hence the term, moral choice blindness; see Johansson et al. 2005); most subjects were also more than happy to provide justifications for the moral beliefs *they didn't even originally claim to have.*

This may not seem to be such bad news, because it mainly demonstrates that people's moral judgments are fickle, and elusive beliefs are unlikely to make a great difference to people's moral behavior, which is what we are ultimately interested in. But in a recent (thus far unpublished) study, Thomas Strandberg and his colleagues found that choice blindness manipulations can indeed lead to lasting changes in evaluative attitudes, especially when people are encouraged to proffer elaborate confabulations for their putative views (Strandberg et al. forthcoming). In this study, people received false feedback about their political views regarding various issues such as gasoline taxes, energy subsidies, or public schooling. It was not only found that detection rates were low; importantly, subjects' attitude shifts in the manipulated direction were still identifiable one week later. In particular, the effect was strongest for subjects who had been assigned to a *Long Confabulation* condition in which they were asked to offer comparatively thorough reasons for their beliefs, suggesting a potentially lasting impact of everyday confabulatory episodes on their judgments. People often confabulate reasons for an externally induced attitude. When their efforts are sincere, and they genuinely believe to be articulating their actual reasoning, they carry their adopted beliefs and the reasons attached to them with them into the future.

We have seen that *post hoc* reasoning and confabulation about practical and moral decisions are far more pervasive than one might intuitively suppose. But this does not have to be bad news. In the following two sections, I will anticipate some of the main points of chapters 2 and 3, which are supposed to show that moral reasoning, even when it comes *after the fact*, is not at all superfluous. First, I will briefly hint at the possibility that an obligation to justify a moral judgment can be shared within a social practice of moral reasoning. This point will be further elaborated in chapter 3. In the last two sections of the present chapter, I wish to briefly revisit the claim that prior moral reasoning can shape subjects' automatic intuitive responses. This will then become the main topic of chapter 2.

10 The Social Structure of Moral Reasoning

Social intuitionism relies upon an important yet rarely acknowledged assumption. Roughly speaking, the SI model takes the effectiveness thesis to hold that subjects are justified in their moral judgments just in case they hold them *because* of the justifying reasons they have. For that to be the case, these reasons must be introspectively accessible to that particular subject. We can call this the accessibility requirement:

Accessibility Requirement (AR)

A (moral) judgment that *M* made by subject *S* counts as being based on reasons only if *S* has internal access to the set of reasons $\{q, r, s, \ldots, n\}$ that justify *M*.

It is a necessary condition for a subject to be rationally entitled to her judgment, the SI model assumes, to have introspective access to the reasons that could justify her judgment. But as the empirical evidence suggests, people have internal access neither to the reasons that justify their judgments nor to their real reasons for making their judgments. From the evidence about the *post hoc* nature of moral reasoning together with AR, the SI model can conclude that in fact, subjects usually aren't justified in holding their moral beliefs. Rather, and in line with the no reasons-interpretation, they hold them on the basis of unjustified intuitive and emotionally charged gut reactions.

One does not, however, have to accept AR as a necessary condition for a subject to be justified in his moral judgments. In fact, we do not demand that the requirement be met in most other areas. Think about scientific beliefs. Generally, we think that one can be justified in believing that the structure of space is non-Euclidian or that there is no highest prime number without being able to explicitly cite the reasons why this is so. We can rely on the social division of knowledge here: there are experts who know about the structure of space and the fundamental principles of arithmetic. By saying that we know that space is non-Euclidian, we are saying that we have good reason to believe that these experts are trustworthy and that they have access to the reasons that are needed to back up these claims. Not all individual subjects need to have access to these reasons themselves.

As Andrew Sneddon argues, we can use this line of thought to construct an explanation for the phenomenon of moral dumbfounding that hinges on the idea that moral reasoning has an essentially *testimonial* structure. Under pressure, people's moral thinking turns out to be fairly "shallow": they do not *really* know why incest is wrong, they just have a strong intuition that it is and try to justify it *post hoc*. The reason why is that in moral reasoning, people tacitly rely on the testimony of experts about moral issues (like philosophers, priests, or politicians): "So long as we can rely on such sources of information, we generally do not need to carry the nuts-and-bolts details of moral issues or theories around with us" (Sneddon 2007, 736). We don't need to have individual access to the intricate details of morality to be justified in our moral judgments. In this picture, we can explain why the phenomenon of moral dumbfounding does not necessarily give rise to a general skepticism about morality and why it does not show that in general, subjects are radically unjustified in their moral beliefs. They are unjustified in the sense that they do not have access to the justifying reasons themselves. But as we have seen, if we drop AR in favor of an externalist picture of moral justification, we can make sense of both the empirical evidence about how people actually reason and the normative view that is maintained by ET, because the effectiveness of moral reasoning does not stay within one mind, as it were, but spreads across a social network of individually achieved and testimonially inherited justification.

11 The Causality Requirement

But the social intuitionist debunking of ET relies on an even more demanding background assumption, which can be called the causality requirement:

Causality Requirement (CR)

A (moral) judgment that *M* made by subject *S* counts as being based on reasons only if conscious consideration of the set of reasons {*q*, *r*, *s*, …, *n*} that justify *M* causes *S* to hold that *M*.

According to the SI model, a moral judgment does not count as being based on genuine moral reasoning if it does not meet CR, the requirement that a judgment be directly caused by episodes of conscious reasoning immediately preceding the judgment. It is not enough merely to have internal access to the reasons that possibly justify one's judgment. Rather, for a particular judgment to be *based on* reasoning in a way that entitles a subject to hold it, the judgment must have been *caused by* the reasoning. (We can see that AR is a precondition for CR, because for conscious consideration of reasons to be possible, they must be internally accessible.) If we apply CR to Haidt's *Incest* scenario, we can see how the judgment that what the siblings did was morally wrong falls short of the requirement. Conscious consideration of the set of reasons {incest is wrong because inbreeding is dangerous, incest is wrong because it jeopardizes family values, …, *n*} cannot have caused the judgment because these reasons do not apply to the scenario (here: *Incest II*), and consciously considering *that* fact does not make people refrain from their judgment.

How can reasons bear an effective connection to a judgment that is supposed to be based on these reasons when conscious consideration of the reasons does not directly cause the judgment? Terry Horgan and

Mark Timmons have developed an interesting suggestion as to how to solve this problem. Moral principles, and moral reasons more generally, can be possessed "morphologically," that is, in a way (a) that reliably conforms to the content of the principle, but (b) such that it need not be consciously represented at the very time of its employment:

> Moral principles may be, and typically are, possessed by morally mature individuals morphologically, and the morphological possession of moral principles then *procedurally guides* the "automatic" formation of particular judgments. Such principles, to be procedurally operative on some occasion, need not be explicitly represented by some occurrent mental state in the individual on the occasion in question. (2007, 287)

In this view, people make their judgments according to habitualized patterns of moral reasoning that are subconsciously activated in a given situation. The reason they can be activated to begin with, however, is that moral reasons did play an effective role in the *acquisition, formation, maintenance, and correction* of these habitualized patterns of judgment and reasoning. In the course of moral education, explicit moral reasons literally become embodied in the emotional-intuitive mechanisms with which people respond to the morally relevant features of a situation. People come to learn that actions of type *X* are wrong and that they are wrong because they generally have the properties $\{q, r, s, \ldots, n\}$ that render them wrong. Obviously, this practice of habitual reasoning is prone to misfire, like every other habit, in situations it has not been tailored to. Uhlmann's study on the racially biased employment of moral principles mentioned above bears witness to this. But, to use an analogy, from the fact that in experimental settings, seagulls can be made to incubate egg-shaped wooden dummies, it does not follow that in ordinary cases, what they do is not really incubating.

Conclusion

ET holds that moral reasoning is effective. It says that the reasons that justify our moral judgments are the reasons that are causally responsible for the fact that we make those judgments. The SI model challenges that view, arguing that in the light of the empirical evidence about how people actually reason, it cannot be maintained. As a matter of empirical fact, people arrive at their moral judgments unconsciously and intuitively. Only the result of that process becomes accessible to conscious awareness, the rest remains an emotionally charged intuitive process. Contrary to ET, episodes of reasoning come after the fact, and bear no effective connection to subjects' judgments. Defendants of ET have a response to this, arguing on conceptual grounds that from the fact that in everyday moral reasoning, AR and CR are not met, it does not follow that people's moral judgments are not based on genuine reasoning. Episodes of genuine reasoning are *embodied* in our habitualized, intuitive responses. The role played by conscious reasoning, then, is to make explicit the normative reasons that (would) justify our judgments and that did play a causally effective role in the acquisition, formation, and maintenance, in short: the education of our moral intuitions. The fact that conscious moral reasoning usually comes *post hoc* in a *temporal* sense does not threaten the core claims of moral rationalism regarding the psychology of moral judgment. Moral reasoning often *is* like Minerva's Owl—it starts its flight when the job is already done. But this does not mean that before it started, we did not get the chance to listen to its advice.

2 Educated Intuitions

Introduction

In his *Philosophy of History*, Hegel writes: "We cannot ever give up thinking; that is how we differ from the animals. There is a thinking in our perception, in our cognition and our intellect, in our drives and our volition (to the extent that these are human)" (Hegel 1988, 10). I wish to argue that he was on to something; rational thought is, indeed, an invariable element even in judgments that seem to be made without any thinking. Moral judgments are based on automatic processes. Moral judgments are based on reason. In this chapter, I argue that both of these claims are true, and show how they can be reconciled.

To some, this might seem like a trivial endeavor. But this is not so. In the previous chapter, I have explained the anti-rationalist trend in recent empirical moral psychology. After decades of rationalist dominance under the auspices of a cognitivist paradigm (Kohlberg 1969), moral psychology underwent an emotionist turn (Haidt 2007; Haidt and Kesebir 2010; Sauer 2011a), and the evidence for the *post hoc* nature of moral reasoning is taken to show that we arrive at moral verdicts on the basis of quick, often emotionally charged intuitions, rather than episodes of controlled reasoning and conscious deliberation (Haidt 2001; Uhlmann et al. 2009).

These developments are part of a more general trend in the psychology of cognition and action to study the pervasive automaticity of human judgment and behavior. It seems that to a surprising extent, judgment formation and action are based on processes that remain largely unconscious (Bargh 1994; Bargh and Chartrand 1999; Dijksterhuis 2004, 2006; Wilson 2002). Often, people do not have access to what really drives their behavior (Sie 2009; Wegner 2002), and are oblivious to what triggers certain judgments or behaviors (Nisbett and Wilson 1977, 1978; Langer, Blank, and Chanowitz 1978). Moral judgments are not, these theorists argue, based on critical reflection and proper weighing of reasons, but on uncontrolled, emotionally charged states of intuitive (dis)approval. Call this the *automaticity challenge*.

In this chapter, I argue that the automaticity challenge to rationalist accounts of moral judgment—which is really just the other side of the anti-rationalist medal—can be met. More precisely, I aim to show that the automaticity challenge rests on problematic, and ultimately mistaken assumptions concerning the nature of automatic mental processes. In a nutshell, I maintain that proponents of the anti-rationalist challenge about moral judgment make a plausible case for the automaticity of moral judgment, but fail to show that there is anything problematic about this automaticity from a rationalist perspective.

Call the claim that a mental process cannot be both automatic and rational, or to put the same point differently, that automaticity excludes rationality, the *incompatibility thesis*. Despite plenty of obvious counterexamples, the thesis is widely accepted.[1] We can undermine this thesis by finding traces of reason in our emotional and intuitive reactions. Since we have no metaphysical guarantee that our feelings and intuitive reactions will live up to that demand, their normative quality will have to be secured indirectly. My suggestion is to start from the notion of a "second nature," and draw on the idea that emotions and intuitions can be *educated*. In the process of moral upbringing, rational grounds become embodied in our intuitive thinking. The following chapter makes a conceptual as well as empirical case for the claim that moral judgments are based on educated intuitions.

This chapter has eleven sections. In section 1, I use Jonathan Haidt's SI model as an example to set out the automaticity challenge in more detail and discuss the most important empirical evidence for the automatic character of moral judgment. I argue that the SI model rests on assumptions that are complementary to the incompatibility thesis, and that it mistakenly ties the rationality of a mental process to the fact that it is or is not conscious. I then discuss recent philosophical attempts to free up conceptual space for the possibility of automatic yet rational forms of cognition, and explain how habits and the concept of a "second nature" figure in those attempts (sections 2, 3, and 4). I show that the incompatibility thesis has been endorsed by many philosophers of action as well, and why this move has seemed attractive to many. But once we see that there is nothing conceptually dubious about automatic yet rational processes—and habits are the prime example here—we can provide empirical evidence for their significance for human cognition in general and moral judgment in particular. Section 5 explains in greater detail the distinction between *post hoc* reasoning and moral confabulation, and shows how the concept of an education of moral intuitions can help draw this distinction (sections 6 and 7). I show how moral education works in general (section 8), distinguish between two different kinds of education of the intuitions (sections 9 and 10), *ex ante* and *ex post* education, and show how educated intuitions can account for the central elements of a normative picture of human moral judgment and agency while leaving the central intuitionist insights intact. The relation moral reasoning bears to our moral judgments is not primarily mediated by episodes of conscious reasoning, but by the acquisition, formation, and maintenance and reflective correction of rationally acquired moral intuitions. I conclude with some remarks about the limitations and possible detrimental effects of conscious reasoning (section 11).

1 The Automaticity Challenge

Let me briefly summarize the results of the last chapter. Simple—and, as I will argue below, overly simple—rationalist models[2] suggest that moral judgment is based on deliberate reflection and the careful weighing of reasons: "one briefly becomes a judge, weighing issues of harm, rights, justice, and fairness, before passing judgment. ... If no condemning evidence is found, no condemnation is issued" (Haidt 2001, 814). Haidt's SI model of moral judgment and reasoning challenges the empirical accuracy of that picture on the basis of two types of evidence:

(i) *Intuitive primacy*. Haidt stresses the fact that people generally arrive at their moral verdicts too quickly for it to be possible to engage in explicit reasoning upfront. Rather, the process of moral judgment formation works much more like perception. People simply "see" whether or not a particular action is morally wrong, and base their judgment on what intuition tells them. This process is often accompanied by quick flashes of emotional (dis) approval (Wheatley and Haidt 2005; Schnall et al. 2008).

(ii) *Moral dumbfounding.* That people engage in explicit moral reasoning is an uncontroversial fact. The SI model, however, questions whether these episodes of conscious deliberation are causally efficacious. Experimental vignettes that trigger a strong moral intuition yet render most possible justifications for moral condemnation pointless have shown that people do not suspend and/or change their moral beliefs if no appropriate reasons can be found. Rather, they enter a state of moral dumbfounding, unable to articulate good reasons for their moral intuitions. A detailed discussion of this evidence was given in the previous chapter.

These findings are taken to suggest that moral judgments are based on automatic System I processes. The intuitive primacy in moral judgment formation is said to give us reason to think that episodes of moral reasoning subjects engage in come *post hoc*. They provide mere rationalizations of the moral intuitions subjects already have. Remember that I have called this the *post hoc* thesis:

Post Hoc Thesis (PHT)

When people engage in moral reasoning, they do so after a moral judgment is reached.

But the SI model, as I explained earlier, makes even stronger claims. Haidt argues that moral reasoning not only comes *post hoc*, but that typically, it is utterly superfluous in bringing about moral judgments (Saltzstein and Kasachkoff 2004; Clarke 2008; Musschenga 2008). The justifying reasons subjects might have play no role whatsoever in the formation of their beliefs (Prinz 2007, 31). On this account, moral reasoning is like confabulation (Hirstein 2005): it does not verbalize the reasons that actually lead subjects to adopt certain judgments. People do not make moral judgments because they reason from legitimate, morally salient considerations and well-founded principles, but rather choose the reasons and principles *post hoc* that best justify the judgment they want to end up with (Uhlmann et al. 2009). I suggested calling this the confabulation thesis:

Confabulation Thesis (CT)

Moral reasoning is a matter of causally ineffective confabulation.

The automaticity challenge—one prime example of which is Haidt's SI model—is supposed to attack the main tenets of rationalism concerning the psychology of moral judgment. Rationalism holds, among other things, that the normative reasons that justify people's moral judgments are also the causally effective explanatory reasons for why they make those judgments. This was referred to as the effectiveness thesis:

Effectiveness Thesis (ET)

The justifying (moral) reasons subjects have for their (moral) judgments figure in true causal explanations for why they hold these judgments.

This is the main claim rationalist models of the psychology of moral judgment are committed to. It is important to carefully distinguish the *post hoc* thesis from the confabulation thesis, because only the latter denies the effectiveness thesis. The former is compatible with the fact that in episodes of *post hoc* reasoning, people make explicit the reasons that really did play an effective role in how they arrived at their assessment of a given scenario, albeit unconsciously. The confabulation thesis denies that this is the case.

The challenge the SI model poses toward the rationalist position is based on the idea that if one can show that moral judgments are largely based on automatic processes (emotional reactions and intuitive cognition), one has *eo ipso* shown that the effectiveness thesis does not hold—which means that the episodes of reasoning people do engage in must be confabulatory, as the confabulation thesis states. Recently it has been shown, however, that this move is not as innocent as it may seem, but only follows on the basis of two further tacitly made assumptions (Sneddon 2007; Horgan and Timmons 2007; Sauer 2011b). For one thing, the SI model holds that unless the accessibility requirement is fulfilled, the effectiveness thesis must be false, see the accessibility requirement:

Accessibility Requirement (AR)

A (moral) judgment that *M* made by subject *S* counts as being based on reason only if *S* has internal access to the set of reasons {*q*, *r*, *s*, ..., *n*} that justify *M*.

Second, and perhaps more importantly, the SI model not only requires that for moral reasons to be causally effective in bringing about a moral judgment, these reasons must be accessible to the judging subject; it also holds that for reasons to be operative in moral judgment formation, conscious reasoning on the part of the judging subject must immediately precede the judgment: the "key part of the definition of reasoning is that it has steps, at least a few of which are performed consciously" (Haidt 2001, 818). We can call this the causality requirement:

Causality Requirement (CR)

A (moral) judgment *M* made by subject *S* counts as being based on reason only if conscious consideration of the set of reasons {*q*, *r*, *s*, ..., *n*} that justify *M* causes *S* to hold *M*.

The empirical evidence mentioned above—the intuitive primacy and the phenomenon of moral dumbfounding—clearly shows that in many cases in which a moral judgment is made, neither the accessibility nor the causality requirement are met.

A closer look reveals that the two requirements are in fact restatements of the incompatibility thesis. They identify the rationality of a process with the accessibility and conscious awareness characteristic of controlled System II processes; conversely, the requirements rule out—by conceptual *fiat* alone—that automatic processes can be based on reason.

The incompatibility thesis is not a trivial assumption anti-rationalist models of moral judgment could easily do without. It is a core commitment that does major work in their argument. In fact, the thesis is so central that Jesse Prinz and Shaun Nichols, for instance, have straightforwardly included it in their definition of moral rationalism, which they characterize negatively as the position that holds that moral judgment can occur "in the absence of ... emotions" (Prinz and Nichols 2010, 116). Emotions are a paradigm example for automatic System I processes, and rationality is defined in terms of the absence of these processes. For this definition to make sense, it must be assumed that a process cannot be both automatic and rational at the same time.

Let me clarify. The incompatibility thesis says that automatic processes cannot be rational. This is a universally quantified claim that holds that for all mental processes *P*, if *P* is automatic, then *P* is not rational. Prinz and Nichols's definition of rationalism, however, has it that rationalism merely entails that *some* moral judgments can be made without the involvement of feeling. And this definition, it seems, is not committed to any universally quantified proposition. This impression is misleading, however: while it is true that Prinz and Nichols's definition allows rationalism to be true even when some moral judgments occur in the presence of emotions, it is committed to the claim that *when* this happens, the resulting moral judgments are not rational. Only those moral judgments that occur in the absence of emotions count as being based on reason. Because emotionism makes the modally strong claim that moral judgments *must* involve emotions, the former claim would be sufficient to establish the truth of rationalism, which is taken to hold that no members of the subset of rationally made moral judgments can be made in the presence of emotion. And this *is* the incompatibility thesis.

I have now briefly rehearsed the main argument of the previous chapter. In what follows, I shall argue that we can account for the effectiveness thesis in a way that does justice to (i) the largely automatic and intuitive character of moral judgment and (ii) the *post hoc* nature of moral reasoning, but (iii) does not entail that moral reasoning is confabulatory and thus causally ineffective. I will argue that the automaticity of moral judgment can be squared with its rationality just in case moral judgments are based on patterns of moral reasoning that have, through a process of moral education, become habitual. I will show why it is legitimate to think of habits—acquired automatic processes—as processes that can be placed in the space of reasons, and I will make good on the claim that moral judgments are in fact based on such educated intuitions.

2 Habits and Practical Reason

Habits are "learned dispositions to repeat past responses" (Wood and Neal 2007, 843). As Bargh and Chartrand observe, habits can be acquired intentionally as well as unintentionally (figure 2.1). Habits are acquired, they argue, through the "frequent and consistent pairing of internal responses with external events" (1999, 468). Both the deliberate execution of a behavioral pattern and its unintentional enactment lead to a process of automatization.

One should understand the concept of a *disposition* used in the above definition of habits as a placeholder for automaticity: habits are behavioral patterns whose execution is triggered in certain circumstances and becomes automatic over time. Habitualization can be plotted as an asymptotic curve, representing the relation between the number of repetitions of a given piece of behavior and its degree of automaticity (Lally et al. 2010, 1002). Figure 2.2 shows an example of how a habit to exercise for fifteen minutes before dinner is formed.

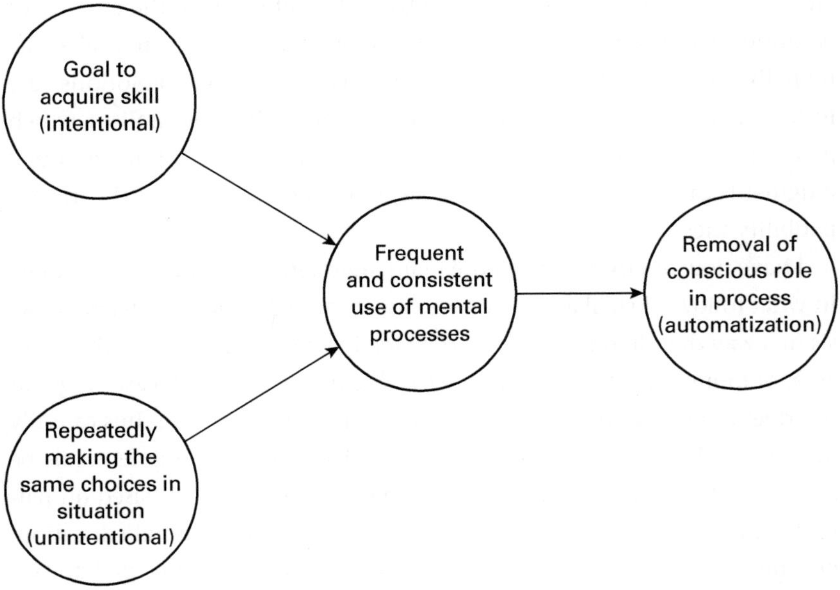

Figure 2.1
Intentional and unintentional routes to the automatization of a psychological process (adapted from Bargh and Chartrand 1999, 469).

The very fact that automaticity is distinctive of habits, however, has led many modern philosophers of action to ignore their significance for a comprehensive theory of practical reason. But this brush-off has not always been so common. Schiller wanted to reconcile the strict Kantian dichotomy between "duty" and "inclination" in his concept of a beautiful soul, a person whose character traits lead her to automatically act in accordance with what is morally required (Schiller 2004). Hegel made the even stronger claim that habituality is *necessary* for moral action, and that explicit deliberation about what to do is (at least potentially) "unethical [*unsittlich*]" (Hegel 1986, 323) because it questions established rules and practices on the basis of an abstract moral point of view. And early twentieth-century philosophical anthropologists (Scheler 2007; Plessner 1928; Gehlen 1940) did not tire of emphasizing how important individual habits and structured, intersubjective habits—conventions and institutions—are for creatures with the degree of cognitive flexibility and plasticity humans enjoy. Recent findings support this insight. It has been established that one of the key benefits of habitual action is its cognitive efficiency: "Habits potentially

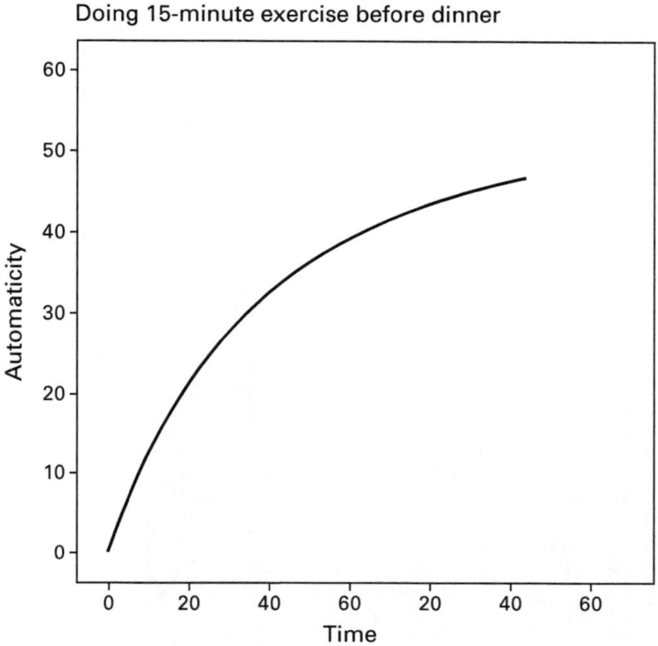

Figure 2.2
Example of increase in automaticity (adapted from Lally et al. 2010, 1002).

free people to engage in other kinds of thoughtful activities such as rumination of past events and planning for future activities" (Wood, Quinn, and Kashy 2002, 1295). It is thus not only inevitable, but also pragmatically rational for agents to rely on habits.

Studies on nonhuman mammals show how the automatization of a given piece of behavior saves expensive cognitive resources (see also Duhigg 2012, 3–31). In one interesting experiment, Ann Graybiel put rats into a T-shaped maze and measured their patterns of neural activity. The rats ran toward the end of a corridor, where they received a reward on either the right or the left side of the top of the T. When the rats had done this repeatedly, their neural firing accumulated in the behavior at the beginning of the task, and toward the end when the reward was received. While they were moving *through* the maze, however, their neural activity decreased significantly (figure 2.3).

Graybiel explains that "habits, whether they are reflected in motor or cognitive activity, typically entail a set of actions, and these action steps

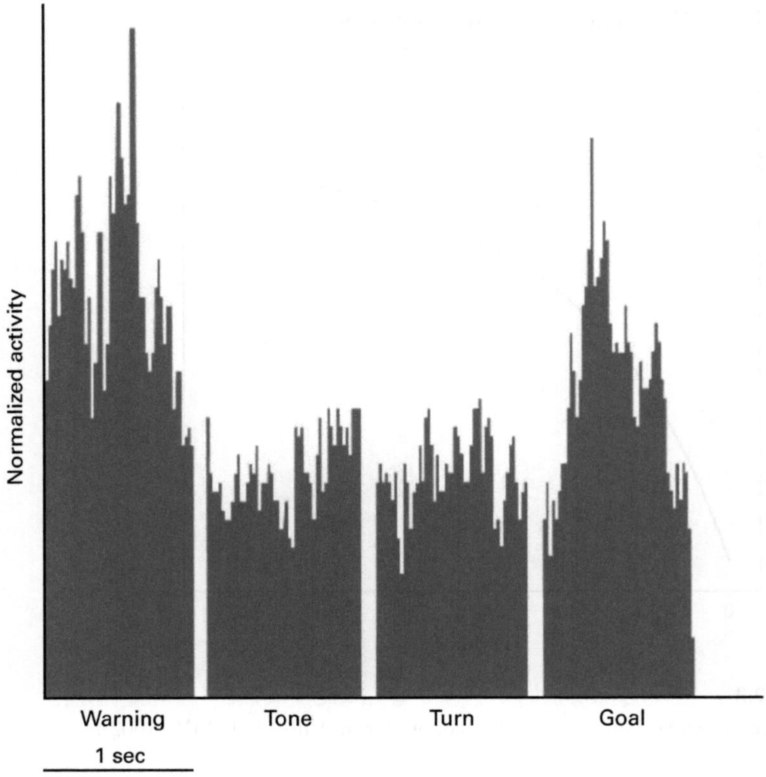

Figure 2.3
Heightened neural responses at action boundaries in macaque monkeys (adapted from Graybiel 2008, 376).

typically are released as an entire behavioral episode once the habit is well engrained" (Graybiel 2008, 375). (The process by which neural activity only remains required at the beginning and end of the enaction of a behavioral episode is called *chunking*.) This, among other things, is why the perception-like character of emotion is so important for moral judgment:[3] in order to "release" an episode of moral judgment formation, the judging subject must first be alerted to the fact that something of moral significance is at stake in a given situation. Emotions pick up on the morally salient features of situations, which then trigger the habitualized moral intuition.

Figure 2.4 provides an illustration of how, once a given piece of cognitive or physical behavior has become automatic, conscious deliberation and reasoning can stop intervening and outsource control to habitual processes.

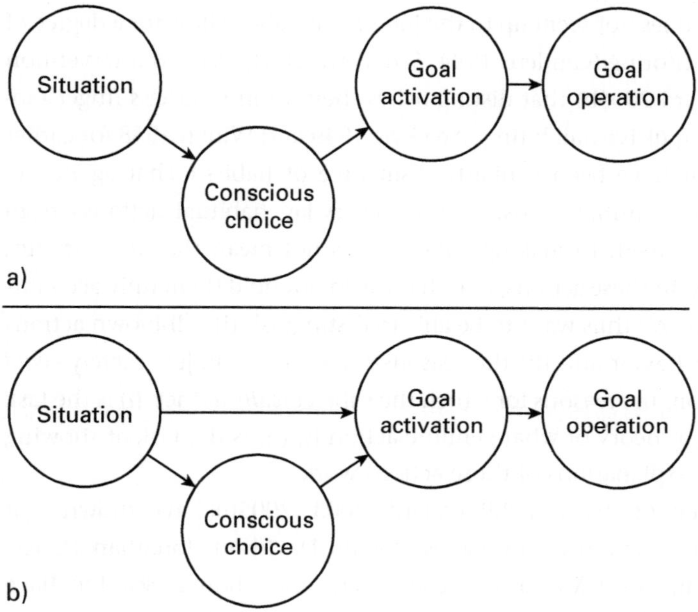

Figure 2.4
(a) Conscious, intentional mediation of goal pursuit within a situation and (b) automatic activation and operation of goals by situational features following repeated choice of the same goal (adapted from Bargh and Chartrand 1999, 470).

To be sure, conscious awareness of the activation of the automatic process is still present; oftentimes, this awareness will reflexively monitor whether the task is carried out well and whether the respective goal has been successfully achieved.

3 Intellectualism and the Reasons Theory

Why is it, then, that habitual action has been so widely ignored in philosophy of action, or that the very possibility of genuine habitual action has been rejected in the first place? The "three categories of phenomena" that, according to David Velleman, "philosophy of action must … account for," are "mere happenings, mere activities, and actions" (2000, 4). Neither the first nor the third category seem suitable for habitual behavior. But the residual category of "mere activities," that consist in a "partial and imperfect exercise of the subject's capacity to make things happen" (Velleman

2000, 4) also does not seem up to the job.[4] Most habits allow for a degree of reflexive monitoring (Giddens 1984, 5–6; Pettit 2001, 39) and intervention control (Pollard 2005b) that distinguishes them from mindless finger tapping and compulsive nail biting. (See Ryle 1949 and Winch 1958 for earlier accounts that have been similarly dismissive of habits.) That agents do not consciously initiate most of their everyday habitual actions—from brushing their teeth to making coffee—does not mean that they are not fully involved in these actions, and that we do not hold them fully accountable for them. We thus want to be able to distinguish (i) full-blown actions from mere behavior and (ii) the reasons for which a subject merely *could have* acted from the reasons for which the subject *really did* act. (i) is the task of providing a theory of what genuine action is, (ii) is the task of showing how reasons-explanations of those actions work.

Joshua Gert (2003) and Bill Pollard (2003, 2005b) have shown that philosophers of action as diverse as Donald Davidson, Jonathan Dancy, Warren Quinn, John McDowell, Joseph Raz, and Thomas Scanlon have tried to achieve the above two tasks by making one and the same assumption, namely, that genuine actions, as opposed to mere mindless behavior, are done for reasons; and that an agent only really does something *for* a reason if she acted "in the light" of that reason.

The way these philosophers spell out this "in the light of" relation subjects need to bear to their reasons is strikingly similar to the accessibility and the causality requirement. Remember that to get their challenge to rationalism about moral judgment off the ground, many empirical moral psychologists assume that for a judgment to be based on reason, conscious consideration (CR) of internally accessible (AR) reasons must bring a subject to accept a certain moral judgment. On the other hand, philosophers of action hold that for an action to be performed "in the light of" reasons, these reasons must somehow exert their motivational and justificatory force upon the acting subject; and the way reasons do this is by being "present to the agent's consciousness" (Dancy 2000, 129) or by "reveal[ing] the favourable light" of the "projected action" to the agent (McDowell 1978, 79). In order "for a consideration to be an operative reason for me," Scanlon writes, "I have to believe it" (Scanlon 1998, 65). Pollard and Gert have coined different names for this requirement, from "judgment thesis" (Gert 2003) to "conception constraint" or simply "reasons theory" (Pollard 2005b) of rational action. According to the reasons theory, task (i) is easily achieved: genuine actions are guided by reasons, whereas mere behavior is not much more than a brutish, mindless response to internal or external stimuli. And the theory has no problem with task (ii), either: an agent's real reasons, and the ones that do not only justify a subject's action from an outside perspective, but explain her actions in terms of what she saw in it, are those that the agent is consciously aware of.

In the psychology of moral judgment, the challenge to the effectiveness thesis is based on the very same intellectualist assumption philosophers of action make: for reasons to be operative in my arriving at a moral verdict, and thus for these reasons to render my judgment rational, I must be aware of them, and consciously consider them in advance. But, so the challenge goes, as an empirical matter of fact, this is not what happens. Most of the time, my moral judgments are based on quick and effortless intuitions (Haidt's "intuitive primacy") and episodes of reasoning come after the fact (the *post hoc* thesis); therefore, reasons are not causally effective in how people arrive at their judgments (the confabulation thesis). It is clear, however, that the automaticity challenge only delivers this anti-rationalist result on the basis of the aforementioned intellectualist assumptions concerning the nature of reasons, and what it means for them to be operative in subjects' judgment and behavior. Luckily, we do not have to accept these assumptions.

4 Reason, Habits, and Second Nature

Can we sketch a theory of action and practical reason that avoids this implausible dismissal of automatic action, leaves room for its rationality, and thus helps to meet the *automaticity challenge* to moral rationalism? Can we develop an alternative picture of how reasons can become effective in moral judgment in a way that efficiently bypasses subjects' conscious awareness?

There are two different ways that this could be done. The direct way would be to argue for cognitivism about automatic processes, preferably emotions, but I resist the temptation to choose this strategy here. In chapters 6 and 7, I will briefly explain why. The indirect way is to pay attention to the malleability of automatic processes: if one can find evidence for complex cognitive processes that used to be executed by System II, but—over the course of habitualization and education—have become automatic over time, we have no reason to think that these processes have changed as far as their rationality is concerned. The idea here is structurally similar to the *parity principle* made prominent in the context of the extended mind hypothesis (Clark 2010). If we would not hesitate to classify, on the basis of its functional characteristics, an event or operation as mental if it went on "in the head," but as it happens, it does not, then we should not hesitate to classify the operation as mental. Similarly, we can propose a *parity principle for automatic processes*:

Parity Principle for Automatic Processes (PPA)

If we would not hesitate, on the basis of its functional characteristics, to call a process "rational" were it performed consciously and effortfully, but, as it happens, it has become habitual and automatic over time, then we should not hesitate to call it "rational."

This principle suggests that there is no reason to think that consciousness is a necessary condition for rationality, or that the two are congruent.

Stretching the concept of rationality so widely to include the notion of automaticity will not do, and it will not convince those who have different conceptual intuitions. Therefore, my argument hinges on the idea that if, as Kahneman and Frederick put it, "complex cognitive operations eventually migrate from System [II] to System [I] as proficiency and skill are acquired" (2002, 51), and if there is empirical evidence that this is the case in the domain of moral judgment and reasoning, then we have no reason to think that simply because the *modus operandi* of a mental process has changed from "controlled" to "automatic," its status as a rational process must have changed as well.[5]

Before I turn to the empirical question how the education of (moral) intuitions works, and which types of education of the intuitions there are, let me briefly discuss the problem of what it is that renders a habit rational from yet another perspective. The rejection of the "reasons theory" of rational action as well as the accessibility and causality requirements on rational moral judgment does not relieve the obligation to account for the difference between action and mere behavior (task [i] from above) and between the reasons that really were effective in bringing about an action from those that merely could have been (task [ii] from above). How can this be done?

5 From *Post Hoc* Reasoning to Confabulation

Bill Pollard has argued that the concept of *Bildung* (education) makes it possible to *naturalize* the space of reasons in a way that allows us to reconcile the habituality of an action with its rationality. It is clear that habitual judgments and actions do not satisfy the accessibility and the causality requirement, because "habitual behaviours ... are automatic, which is to say that they do not seem to be preceded by deliberation of any sort" (Pollard 2005a, 74). But how can there be reasons in action if the requirements of the "reasons theory" aren't met? How can reasons be operative, if they are not in any way present to the subject? In my idiom: how can the effectiveness thesis hold if AR and CR are not satisfied? There are two alternative ways to solve this problem.

A first way to explain away this tension in the concept of habitual action would be this: habitual action does not draw on controlled cognition. To a large extent, the actualization and execution of habits goes on at a subconscious level that is typically not accessed by the acting subject. This does not mean, however, that the agent's practical reasons cannot be present *on this subconscious level*. By that account, subjects can act on reasons in ways that are not mediated by conscious awareness and explicit deliberation. But this is not to say that these play no role at all:

> On the contrary, her reasons could be brought to consciousness were she, or somebody else, to enquire about why she did what she did. The key thought is that her reasons are, as it were, already in place when she acts, and are thereby ready to be discovered afterwards should the need arise. On this conception of reason giving, success is marked by these hitherto subconscious states being made explicit. (Pollard 2005a, 79–80)

The above quote makes clear why, as I have argued above, we must carefully distinguish the *post hoc* from the *confabulation thesis*. Episodes of *post hoc* reasoning need not be confabulatory. In most cases, we have no reason whatsoever to think that when people give reasons for their moral beliefs, they are confabulating. Despite the *post hoc* character of moral reasoning, genuine moral reasons are effective in how people arrive at their verdicts: they figure effectively in the acquisition, formation, and maintenance (that is, the education) of subjects' moral intuitions, and make a psychologically real difference to people's moral beliefs. Effective moral reasoning requires nothing more than this.

A second way to explain the effectiveness of reasons in habitual action is this: we can show, the suggestion goes, that an action is rational if the acting subject can come up with a narrative that reconstructs how a given piece of behavior fits into the agent's overall world view and character. What decides whether an action is based on reasons or not, then, is not whether these putative reasons have always already been in place, albeit subconsciously, but whether the behavior in question can plausibly be made to fit into and cohere with the agent's overall system of desires, intentions, goals, and values (Pollard 2005a, 80). But this second option, I shall argue, is not what we are looking for.

6 Rational Habits: The Goal-Dependency of Education

Which of the two solutions is the correct one? Pollard seems to think that both options are equally legitimate. But this is not the case, and the reason for this is that the second solution does not do the job it has been hired for: it conflates the distinction between *post hoc* reasoning and confabulation, and makes episodes of explicit reason giving that are accurate indistinguishable from cases in which subjects merely come up with a "coherent fiction" (Snow 2006, 559). Social psychologists have shown that sometimes, we construct stories that seem to make sense and come up with reasons for our responses which, though plausible, are demonstrably false (Nisbett and Wilson 1977, 1978). A version of the first solution must be true, if such a thing as genuine, nonconfabulatory *post hoc* reasoning is to be possible. At this point, it remains obscure what it means that reasons are subconsciously present at the time a subject acts, waiting to be made explicit. Perhaps not surprisingly, the concept of education sheds light on this issue as well.

Remember that habits are behavioral patterns that become, through repeated enactment, routinized over time. The PPA proposed above suggests that this process is rationality-preserving, because what changes in this process of automatization is not the *nature* of the process, but merely its *modus operandi*. But why is it that agents acquire some habits and not others? Is this a matter of mere luck? And, perhaps more importantly, why is not completely random which habits are acquired and which are not? Consider this example: I am riding home from work on my bike, and I do so, as it were, on autopilot. My unlocking the bike, my leaving the lot, my using the handle bar, are all entirely automatic. But, of course, this sequence of automatic actions is not pointless, and it is not irresponsive to the subtle environmental features that change every day. Rather, these atomic actions all serve my molecular goal of arriving at home. In fact, that I have this goal is why I have developed this particular sequence of habitual actions in the first place: "The reason that these automatic, habitual actions are performed is to serve the agent's chronically accessible goals.[6] Thus, habitual, automatized goal-dependent actions are purposive. The agent's reason for acting—to serve a chronic goal—is not present to her consciousness at the time of acting. Nevertheless, it is operative in her psychological economy. It is a motivating factor that explains her actions" (Snow 2006, 552; see also Snow 2010).

In the course of an agent's education, her practical reasons become embodied in her automatic judgmental and behavioral responses. These reasons are thus both internal to a subject's psychology and external to her conscious awareness and initiation control at a given point in time. Making explicit the reasons that brought me to adopt my after work routine—namely that I want to go home—is an enterprise that comes entirely *post hoc*. But it need not be confabulatory; indeed, it would be a stretch to suppose so.

7 Varieties of *Post Hoc* Reasoning

In meeting the automaticity challenge against rationalism about the psychology of moral judgment, the rationalist must explain why there is nothing intrinsically dubious about *after the fact* justifications. But she must be careful to avoid saying that there is nothing even potentially dubious about such justifications. Sometimes, people irrationally hold on to their convictions; sometimes, reasons are just rationalizations; and sometimes, people are guilty of confabulating.

This distinction is important, and it must be drawn as neatly as possible. The idea that moral intuitions can be educated seems to allow for reasons to become effective somewhere more "upstream": it is alright, I have argued, for intuition rather than reasoning to be the proximate cause of moral judgment, as long as the reasons one cites after an intuition is arrived at did figure in the acquisition, formation, and maintenance of that intuition. But the more "upstream" these reasons become causally efficacious, the more the distinction between genuine *post hoc* reasoning and mere confabulation is blurred. This shows that more needs to be said about the distinction, and it also shows that the concept of education alone, though important, is not sufficient to draw the line between "good" *post hoc* reasoning—which cites the reasons that helped educate your intuitions—and bad confabulation—which cites reasons that are entirely disconnected from the causal genesis of your judgments.

At this point, I want to make a further distinction I have ignored so far: the distinction between *post hoc* reasoning, confabulatory reasoning, and biased reasoning (Liao 2011). *Confabulation* is demonstrably inaccurate, tendentious after the fact reasoning. *Biased* reasoning is tendentious after the fact reasoning, but need not be confabulatory: episodes of biased reasoning typically consist in a partisan selection of relevant and accurate reasons. Mere *post hoc* reasoning need not even be biased. It simply comes after the fact, when the reasons for an intuitively compelling judgment are made explicit.

Moreover, there are two different kinds of causally ineffective *post hoc* reasoning, and it is an intricate question at what point they become confabulatory. Either, one comes up with reasons that really do justify one's judgment (they are good reasons) yet these didn't play a causal role. You used to think that torture can, under certain circumstances, be right. Now you've changed your mind—you think it is always wrong to torture people because it violates their dignity as persons. Let's suppose, for the sake of the argument, that this judgment is correct, and it is correct for precisely this reason. But in fact, this good *normative* reason was not the effective *motivating* reason for your change of mind. Rather, there is this woman you fell in love with who is very strongly against torture, and you started to share her opinion out of affection. This causal path led you to the correct belief and reasoning—that torture is wrong because {*q*, *r*, *s*, ..., *n*}—by accident.

Or, and this is the second kind, one comes up with a bunch of reasons that weren't the genuine causes of your judgment either, but this time the reasons are bad—they are fallacious, or factually inaccurate. You believe that cannibalism is wrong because it is unnatural and offensive to God. But not only are these reasons bad, they are also not causally responsible for why you believe that cannibalism is wrong. You just have this spontaneous response of disgust and horror toward it that you are trying to rationalize after the fact. Should we treat these two cases differently? Or are they equally confabulatory?

It seems to me, perhaps somewhat unsatisfyingly, that whether the former type—the one that is illustrated by the torture example—ought to be considered confabulation is a matter of degree. This does not mean, however, that there are no criteria for this degree: the criterion I would suggest is how tenacious you are in your willingness to confabulate reasons, and how willing you are to let yourself be driven into a state of moral dumbfounding and defend your position nonetheless. People differ with respect to how willing they are to justify hopeless beliefs, or with respect to how happy they are to give up their judgment when exposed to the uncoercive coercion of the better argument. What about people who cite good reasons for their otherwise emotionally driven intuitions and are thus "confabulating" in the first round, but are happy to revise their judgment in the light of undermining reasons that are put forward to them in the second? If they were merely confabulating, and their reasons didn't play any causal role *whatsoever*, how could the fact that those allegedly confabulated reasons have been undermined cause them to change their point of view?

This suggests that there is a second way for reasons to become effective: the first one lies "upstream," and recruits our capacity to educate our intuitions. The second one lies "downstream," and depends on how subjects *would* react—counterfactually—to legitimate challenges to their intuitions and reasoning. On that account, whether one is confabulating is not only a question of whether one cites the reasons that played a formative role for his or her intuitions. It is also a question of whether one treats one's reasons, once they are on the table, *as if* they did play such a role, and whether one is willing to reconsider one's judgment when one's reasons are debunked. If your good, but merely co-opted reasons function in such a reason-responsive way, you are not confabulating, because in that case, there is a real causal connection between your judgment and your reasoning, albeit an indirect and delayed one. If your co-opted reasons aren't like that, however, then you *are* confabulating, because this reveals that there really was nothing but, say, your affection to the woman that made you change your mind, and this affection was not coupled (not even counterfactually) with any rational insight. Moral reasons are free-floating in that sense: one can grab hold of them if one has to, and even though they might have played no significant role at first, they might do so in the future, when it comes to defending one's position in a moral conversation, or to altering it in light of a new case.

8 Moral Education: Experience and Teaching

Moral intuitions are, like other kinds of judgment and behavior, educated through experience and teaching (Hogarth 2001).

The first of the two ways—experience—is nicely reflected by Kohlberg's stages of moral development (1969; see also Darley and Schultz 1990). These do not only correspond to internal cognitive developments on behalf of the subject, but to external changes in the social environment subjects are confronted with. It is hardly a coincidence that the patterns of reasons subjects come to master over the course of their moral development tend to mirror the social, interactive environment subjects typically find themselves in and most depend on. Children's first interactive context is their parents. Accordingly, the "preconventional" reasons they put forward for their judgments refer to authority and punishment. As peers and friends become more important, their moral reasoning typically refers to the "conventional" rules of a specific community. When adolescents have developed a stable identity of their own, and start to endorse values and norms independently of any particular social context, their moral reasoning refers to "postconventional" universal rights and abstract norms of justice and fairness. A crucial part of this education through experience consists in joint action with other people, which is impossible without a shared background of moral norms. Accordingly, feedback from unsuccessful, interrupted collective action will have an influence on the moral norms one is equipped with in future attempts to act jointly with other people. To put it in the form of a slogan: "Moral stages are not structures of thought. They are structures of action encoded in thought" (Reed 2008, 373). Experience is the process by which structures of action become encoded in our intuitive responses.

This provides a hint as to how Kohlberg's rationalist moral psychology can be reconciled with Haidt's anti-rationalist position: remember that Haidt's objection to Kohlberg's model was that the latter focused on the explicit justifications offered by his subjects; according to Haidt, the possibility that these justifications might not be causally responsible for people's judgments is not considered. I think he has a point here. But if, as my argument suggests, not all *post hoc* reasoning has to be confabulatory, and if explicit reasoning can, through experience and teaching, migrate into people's intuitive responses over time—become "encoded" in them—then Haidt and Kohlberg can both be correct. A large amount of moral reasoning comes after the fact, sometimes in the form of confabulation, sometimes in the form of people making explicit the reasons that figured in the *acquisition, formation, maintenance, and reflective correction* of their moral intuitions. And the development of people's reasoning follows Kohlberg's stages. These two views do not exclude each other.

Explicit teaching is another profoundly important means of educating people's capacity to make moral judgments. In fact, it is part of what makes children understand what renders a judgment moral that they be made familiar with different categories of reasons that bear on the validity of different kinds of (social) norms, and why their transgression is prohibited. In a series of studies, Judith Smetana (1984, 1989) found that children learn the distinction between moral and conventional norms on the basis of the different kinds of reasons their parents supply when either norms are violated. Parents tend to refer to considerations of social order and abstract rules in the case of conventional transgressions; in the moral case, parents will request children to take the perspective of others, think about other children's needs and feelings, or refer to considerations that pertain to their rights and entitlements. Children thereby learn which norms it is appropriate to associate their moral emotions with.

Darley and Schultz argue that the acquisition of intuitive knowledge of moral rules and principles frequently depends on various forms of social transmission. In every culture, there are authoritative figures who are licensed to initiate learners into the specifics of a moral code. Over the course of this process, children learn what rules there are, when and whether these rules are open to exceptions, and how to distinguish moral rules from rules designed to protect children against physical danger. Children also learn that moral systems are negotiable, and that this process requires them to understand the *challenge and response* structure of moral reasoning (about which I will have more to say in the next chapter). Regardless of whether this happens in real or role-play situations, children learn how to interpret the disapproval of others as a sign indicating they are being accused of a normative transgression; they learn when it is in order to justify their action, when to excuse it, invoke mitigating circumstances, or flat out deny that any transgressions occurred at all (Darley and Schultz 1990, 541).

Like any other kind of education, the education of (moral) intuitions is a process of habitualization. The level of automaticity with which an intuitive judgmental response is triggered increases with the number of repetitions, a process that consists in a "migration" of controlled and effortful cognitive processes into an agent's effortless, perception-like intuitive system. Dreyfus and Dreyfus's model of intuitive skill acquisition contains an apt description of the stages this migration typically involves (1986, 1991; see also Musschenga 2009). For the *novice* (stage 1) who learns how to perform a task (here: of moral judgment formation), the elements of this task need to be decomposed such that he can become familiar with them. This is what parents do when they teach their children what they ought and ought not do, and why this is the case. The following stages—from *advanced beginner* (2) to *competence* (3)—involve an increase in what might be called *normative automaticity*: the agent not only acquires higher levels of automaticity in dealing with certain tasks, but manages to perform them with greater reliability and a more autonomous and flexible understanding of the subject. Competent subjects have acquired mastery of moral concepts and implicit knowledge of the reasons that count in the context of moral discourse. They are in a position to teach the practice of moral judgment to novices by example. At the level of *proficiency* (4) and *expertise* (5), the agent need not explicitly (and certainly not rigidly) follow the rules she has been taught, but has acquired a perception-like intuitive ability to evaluate which response a particular situation calls for. A proficient moral judge will be more original, creative, independent and reflective in his application of moral concepts and have the ability to improve his or her web of moral beliefs from within. An expert moral judge is a proficient moral judge with metaknowledge about normative and metaethical theories concerning the nature of the practice of moral judgment and reasoning she engages in.

To be sure, norms are rarely taught in explicit propositional form, which means that there must be some mechanism by which subjects, and especially children, can rapidly pick up and store implicitly available normative information. It has been suggested that there is a dedicated *acquisition mechanism*, which can be distinguished from the more motivation- and output-oriented *execution mechanism*, on the basis of which subjects come to grasp salient norms in their social environment (Sripada and Stich 2006; Fitzpatrick 2014). This information is stored in an internal norm database and can be applied in judgments about third parties or, when coupled with the appropriate affective input, as a motivational link to first-personal action. The crucial thing to emphasize is that this mechanism must be able to operate on the basis of scant evidence in the form of outward behavioral cues, from which it can extract the content of the respective norm so enacted (figure 2.5).

It is still an open question whether there are innate constraints on which norms can and cannot be acquired or whether the "Pac Man" thesis—according to which *all and only* the norms present in an agent's environment will be acquired—is true after all. Some normative knowledge may be straightforwardly innate; or, there may be an innately restricted set of possibilities regarding which norms can and which cannot be acquired: a universal moral grammar that allows agents to classify events in terms of causes, intention, means, ends, side-effects, and the basic deontic vocabulary necessary to form moral judgments about these; something like the "big three" CAD model (Rozin et al. 1999), Fiske's four *relational models* account (Rai and Fiske 2011) or Haidt's (2012) five/six *moral foundations* theory may capture the innate lines within which experience then colors;

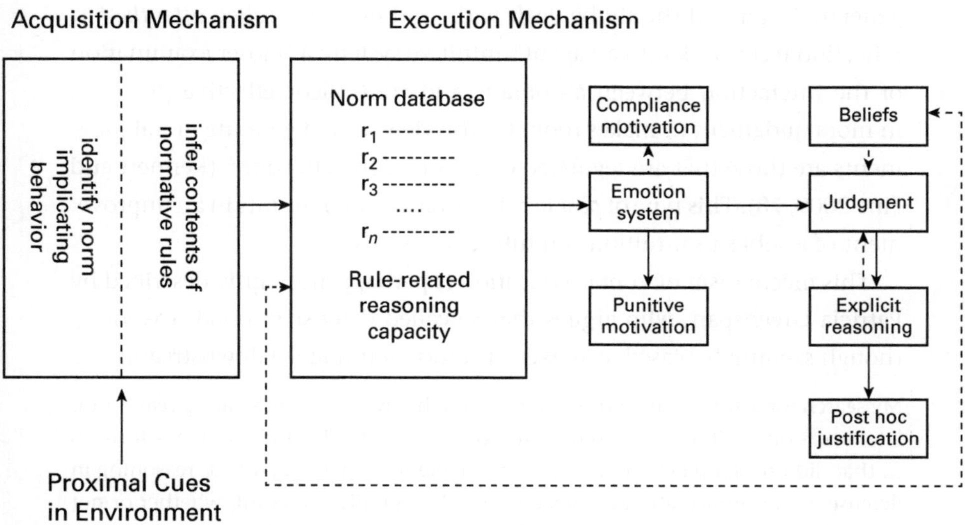

Figure 2.5
A schematic representation of the mechanism of norm acquisition (adapted from Sripada and Stich 2006, 298).

or, finally, the norm acquisition mechanism may be shaped by innate biases that affect which norms present in an environment are more likely to be handed down through the generations (e.g., more disgusting, simple, or otherwise appealing ones; for a more detailed discussion of some of these options, see Sripada and Stich 2006, 297ff.).

9 *Ex Ante* Education

The education of the intuitions is supposed to improve the quality of our intuitive responses to morally salient scenarios. For the most part, rationalist accounts of moral judgment have focused on how reflection and deliberation can regulate our moral emotions and intuitions after the fact. But intuitive education is possible *ex post* and *ex ante*: the latter type is antecedent focused, the former response focused. *Ex ante* education is concerned with the conditions under which a moral intuition is generated:

> Prior reasoning can determine the sorts of output that emerge from intuitive systems. This can happen through shifts in cognitive appraisal, as well as through conscious decisions as to what situations to expose oneself to. In both of these regards, prior controlled processes partially determine which fast, unconscious, and automatic intuitions emerge. (Pizarro and Bloom 2003, 194; see also Gross 2002, 282)

Ex post education is concerned with how an intuition, once it has been generated, can and should be dealt with, and how controlled *after the fact* reflection feeds back into an agent's intuitive system: a "closer examination of the interaction between automatic and controlled reflective processes in moral judgment ... makes room for the view ... that genuine moral judgments are those that are regulated or endorsed by reflection" (Kennett and Fine 2009, 78). This type of *post hoc* deliberation often results in an improvement of a subject's intuitions on future occasions.

This mechanism of *ex ante* education is perhaps most aptly described by Patricia Greenspan, who argues that so-called defensive moral reasoning, though seemingly biased, can exert its rational influence downstream:

> My search for justifying reasons in one case might give rise to motivating reasons on later occasions, if it leads me to react to—or even to identify and consider—features ... that did not stand behind my present judgment. If so, then *post hoc* reasoning in defense of our immediate reactions need not be causally irrelevant, whether or not it reveals their causes on the occasion in question. Indeed, some of our current reactions may themselves have developed from past instances of "defensive" reasoning. (2015, 113)

How exactly does this work? Subjects can influence their moral intuitions *ex ante* by selecting the situational input they are confronted with (Pizarro and Bloom 2003). Situation selection and input control can be very general and far-ranging: many academics, I suppose, have the lingering suspicion that a successful career in business will require them to get their hands dirty, and that it might alter their character in unwanted ways. If you join the mafia, you know that things could get ugly. Avoiding situations one can expect to trigger unwanted intuitive responses gets to the root of problems like these, because relevant factual knowledge often does not suffice to eliminate automatic behavioral responses, whether they are character traits, stereotypes, prejudices, or racial biases.

Input control does not have to be negative, and consist in the avoidance of certain unwanted situations, but can also be about selectively exposing oneself to wanted situational stimuli. People who want to become vegetarians often start reading about the horrors of factory farming. People who register, and want to get rid of, their unwanted racist attitudes can deliberately engage with people of a different race, a strategy that is known to be very effective (Brandt 1979). Monteith and colleagues (2005) have studied, for example, the various strategies people employ to monitor and influence their own racial biases and prejudices. They have shown that under suitable circumstances, people can be very good at registering so-called "should/would discrepancies" in their automatic responses to racially significant stimuli. Subjects can establish—via retrospective and prospective reflection—automatically activated cues that serve to control and help inhibit unwanted judgmental responses, and thus bridge the gap between how people think they *would* and *should* behave.

Ex ante education of moral intuitions need not be focused on external conditions. There are ways to directly influence the formation of intuitive responses by altering the internal processes underlying judgment formation. One striking example for this comes from research on so-called "implementation intentions" (Gollwitzer 1999; Gallo et al. 2009). These *if-then* plans to respond in certain ways upon encountering anticipated stimuli can create "instant habits" (Gollwitzer 1999, 499). Gollwitzer found that implementation intentions can significantly shape people's automatic emotional reactivity in desired ways. In one study, a group of participants managed to reduce automatic responses of disgust and fear toward external stimuli by forming an if-then intention to stay calm and relaxed when confronted with fear-inducing images.

Arguably the most spectacular evidence for how subjects can educate their intuitive judgments comes from research on social prejudice and stereotype activation. Rudman and her colleagues (2001) showed that repeated exposure to suitable stimuli can dramatically influence people's implicit as well as explicit racial prejudices and stereotypes. After taking a class about social prejudice taught by an African American professor, students were less likely to automatically associate typical "black" names with negative concepts such as *laziness* or *hostility*. And this is not a peculiarity only a bunch of scientists are interested in: a quick look at the history of the last two hundred years tells us that mechanisms like these have had a profound impact on our society, and helped do away with many—though obviously not all—irrational responses, ill-founded prejudices, and intolerance.

10 *Ex Post* Education

Ex post education of moral intuitions recruits the human capacity for metacognition: the ability to reflectively monitor one's own cognitive operations and alter them according to standards of rationality or reliability deemed appropriate by the reflecting subject. This is not so much an empirical hypothesis—although it is also that—as a constraint on the concept of a moral judgment: moral judgments do not count as genuine ones unless they are responsive to episodes of rational reflection (Jones 2006; Fine 2006; Sauer 2011a; this idea is developed further in chapters 3 and 6). These episodes play an important role in the education of our emotionally charged moral intuitions to the extent that they feed back into subjects' intuitions, and thereby exert a corrective influence.

The idea that episodes of moral reflection feed back into our intuitive responses and help educate them undermines the incompatibility thesis from yet another angle. Often in the debate between sentimentalists and intuitionists about moral judgment on the one side, and rationalists on the other, "there is no further discussion of the extent to which moral intuitions are amenable to modification as the result of reflection. The … clear focus on the idea that moral judgments are driven by *either* one system or the other, and the need for reflective processes to override intuitive responses, suggests that the scope of reflective modification of moral intuitions is assumed to be minimal" (Craigie 2011, 60; see also Besser-Jones 2012). But this is not so, as the empirical evidence shows.

People use their capacity to make intuitive judgments in utterly different areas—from moral issues to assessments of probabilities, logical relations, and judgments about their own well-being. As diverse as these areas are, the mechanisms subjects recruit to make judgments are often quite similar. The influence of *post hoc* reflection on these mechanisms is well documented, especially in cases where people correct for so-called "mental contamination" (Wilson and Brekke 1994). Schwartz and Clore (1983) have shown that people are both willing and able to take into account irrelevant circumstantial factors that might distort their judgment and to discount them if necessary. In one of their experiments, subjects were asked to evaluate their quality of life under different conditions (on a sunny or rainy day, respectively). When they were given information about how the weather might have affected their evaluations, subjects discounted it, trying to counterbalance the extraneous influence of a gray sky. Now the magnitude of this effect may not be very large, especially when subjects are not given useful additional information about the strength and/or direction of the unwanted influence they might want to discount; when conditions are favorable, however, and subjects are both able and motivated to do so, some biases can be discounted both accurately and entirely (Pinillos et al. 2011).

The habitualization of judgmental patterns in the wake of *ex post* reflection is not a pious hope, but results almost inevitably from the repeated execution of moral reflection:

> The more frequently people perform a behavior, the more habitual and automatic it becomes, requiring little effort or conscious attention. One of the most enduring lessons of social psychology is that behavior change often precedes changes in attitudes and feelings. Changing our behavior to match our conscious conception of ourselves is thus a good way to bring about changes in the adaptive unconscious. (Wilson 2002, 212)

Ex post moral reasoning exerts a rational pressure on subjects to modify or mollify their moral intuitions in accordance with the reasons that become available them, or to give up their intuitions if there aren't any. The perception of should/would discrepancies, the motivation to overcome them, and the evidence that those rational "shoulds" do become effective in shaping a person's automatic responses are prime examples. These processes cannot but have an influence on a person's moral mind-set, however subtle, mediated, and delayed it may be:

It is a general principle in Psychology that consciousness deserts all processes where it can no longer be of use. ... We grow unconscious of every feeling which is useless as a sign to lead us to our ends, and where one sign will suffice others drop out, and that one remains, to work alone. (James [1890] 1950, 496)

For those who still remain unconvinced by this reassurance, here is more evidence for the rational malleability of automatic processes of moral judgment formation. The phenomenon of moral dumbfounding seems to suggest that a fair amount of moral judgments are based on irrational automatic intuitions subjects aren't able to justify. But, as Neil Levy observes, this is just not true; subjects who have a better education *are* able to do so, or do not attempt to if unable:

Haidt's work on moral dumbfounding ... actually demonstrates that dumbfounding is in inverse proportion to socio-economic status (SES) of subjects. ... Higher SES subjects differ from lower not in the moral theories they appeal to, but in the content of their moral responses. In particular, higher SES subjects are far less likely to find victimless transgressions—disgusting or taboo actions—morally wrong than lower. (Levy 2007, 307; see also Haidt, Koller, and Dias 1993)

The only explanation for this seems to be that "the differential responses of higher and lower SES subjects demonstrate that in fact moral intuitions are amenable to education. The greater the length of formal education, the less the likelihood of subjects holding that victimless transgressions are morally wrong" (307). The model developed here can explain that fact; the SI model cannot.

It is striking how grossly this fact is often misinterpreted. Haidt writes about the results that were obtained by running his moral reasoning experiments in cross-cultural and cross-hierarchical settings: "In other words, well-educated people in all three cities [Philadelphia, Porto Alegre, and Recife] were more similar to each other than they were to their lower class neighbors" (Haidt 2012, 22). Haidt takes this as evidence for the cultural variability of our moral gut reactions. But this is very misleading: it is hard to conceive of evidence that more clearly demonstrates the impact of rational education on subjects' intuitions and reasoning, and the convergence that results from it.

Cross-cultural studies in general are a suitable test case for the *Educated Intuitions* model. In particular, it predicts that epistemic factors should play an important role for the quality of people's moral intuitions, regardless of what other cultural differences there may be. Peter Blake and his colleagues (2015), for instance, found that early onset *disadvantageous inequity aversion*—the tendency to disapprove of inequitable outcomes even when such disapproval is against one's interests—is a cultural outlier that, out of the several Western and non-Western countries they investigated, they could only find in children and adolescents from the United States, Canada, and Uganda. Importantly, the results found for subjects from Uganda suggest that the effect is due to epistemic factors rather than mere ethnocentric bias.

A quick look back at the phenomenon of moral choice blindness discussed earlier provides further support for the account developed here. In Hall's (2012, 4) "sleight of hand" study, a surprisingly high amount of people fail to detect the change in the content of the moral statements they are requested to rate their agreement with. However, this pattern did not hold as unambiguously for *politically active participants*, suggesting that people whose intuitions about a subject matter are more deeply entrenched are more likely to have firm moral beliefs and thus less likely to confabulate.

The very idea that emotionally charged intuitive judgments would have to be mere gut reactions, embodied in inflexible "flashes" of disgust or "alarm-like" responses of distress, severely underestimates the significance of affect and intuition for intelligently navigating the natural and social world. It also misrepresents the way they work: for in order to perform this function of attuning an organism to the world such that it can gather and process practically relevant information, emotion and automatic cognition need to be far more experience-based and plastic than anti-rationalists like to suggest (Railton 2014). And indeed, what seem like brute intuitions are often more plausibly construed as a tacit sensitivity to relevant features of the situation. Subjects have difficulty articulating their responses because they rarely have to, at least outside of highly unusual experimental situations designed by savvy researchers determined to show the feebleness of conscious reasoning.

Consider the notion of *risk*. Most people recognize the moral relevance of risk, yet have difficulty pointing out that in Haidt's *Incest* vignette, the fact that the story stipulates Julie and Mark's relationship wasn't harmed by their actions does not entail that what they did wasn't irresponsibly risky.

Their relationship *could* easily have been harmed, and there was no way for them to know that it would not be. When primed for the difference between risk and resultant harm—for instance by an analogous vignette in which a pair of siblings decide to play Russian roulette and survive (Railton 2014, 847)—people find it less difficult to come up with this objection. Presumably, they originally fail to do so because its relevance is further obfuscated by an experimental devil's advocate who insists that only actual harm counts.

Recent evidence on dumbfounding and imaginative resistance confirms the suspicion that many cases of alleged dumbfounding are better described as an automatic resistance to prematurely distrust one's acquired intuitive appreciation of important, though tacit and hard to verbalize, considerations. As Royzman, Kim, and Leeman (2015) have shown, dumbfounding is often best explained by subjects' unwillingness to *believe* the details of a story, either fully or in part, rather than the imperviousness of their intuitions to rational reflection. Despite the fact that they failed to be convinced by the alleged harmlessness of the siblings' actions, participants could nevertheless be brought to explicitly disavow their reasoning for their disapproval of the described action, even when their reasoning had cited potential harm ("Because it *could have* harmed them emotionally/psychologically"; Royzman, Kim, and Leeman 2015, 304; emphasis added). Moreover, more subjects found reasons pertaining to risk more plausible than any of the others. Many subjects then went on to *recant* their abandoning risk-based reasons in response to the devil's advocate's counterarguments, explaining their actions in terms of confusion, feeling pressured, or simply playing along and going through the motions.

The dark side of this tendency is that when subjects strongly believe an action to be morally wrong, they often invent corresponding victims. This has to do with the dyadic structure of moral cognition that construes the world in terms of agents and patients (Gray and Wegner 2009), so that when an agent performs an action that seems severely wrong, the required patients are filled in no matter what (Gray, Schein, and Ward 2014). On the other hand, this shows that rationalization cannot always be blamed on conscious reasoning. Intuition can confabulate, too: harms are often invented and victims fabricated by automatic processes as much as controlled ones.

Moral judgments, I have argued, are typically made intuitively. But moral reasoning, even of the explicit kind, is typically performed habitually, too. Haidt exploits this latter fact in order to produce the phenomenon of moral dumbfounding: if subjects were not used to putting forward certain reasons in support of their verdicts in ordinary situations, they would not attempt to do so—misguidedly—in the extraordinary cases they are given in his experiment. This shows that the education of moral intuitions has two different objects. It is not only particular intuitions with a particular morally salient content—wrongdoing ought to be punished, promises must not be broken, honesty is a virtue—that are being acquired over the course of a subject's moral education, but always also a particular set of reasons that bear on those intuitions. In Haidt's *Incest* case, subjects have a particular moral intuition: *incest is wrong*. But they are also equipped with a set of reasons they have learned to cite as considerations relevant for this very judgment: the harmfulness of inbreeding, the value of family relations, and so forth. Moral education is about both: an improvement of one's intuitions as well as one's appreciation of the reasons one has for them.

The educational processes just described, especially those of *ex post* reflection, can be performed monologically or dialogically, that is, either by interacting with and talking to fellow moral reasoners, or by interacting with and talking to oneself. The SI model misrepresents the nature of intersubjective moral reasoning when it describes it as a process of two or more parties reciprocally persuading and being persuaded by one another. Quite the contrary: Lapsley and Narvaez put great emphasis on how episodes of intersubjective reasoning shape moral intuitions, which they call "chronically accessible cognitive-affective moral schemas" (2004, 24). They argue that as children interact and communicate with the moral "experts" around them—usually their parents—they develop ever more sophisticated abilities of intuitive moral perception. These episodes of dialogical moral reasoning then become integrated into children's identity, and "enable children to organize events into personally relevant autobiographical memories, which provides, in the process, as part of the self-narrative, action-guiding scripts ... that become overlearned, frequently practiced, routine, habitual, and automatic" (26). Such dialogical moral reasoning, about which I will have more to say in the next chapter, is not only a genuine source of improvement of one's moral intuitions. It is a substantive demand of morality to maintain the conditions under which it is possible.

11 Reason and Its Limits

Let me sum up. Our moral intuitions are educated by different *means*: experience and teaching. There are different *types* of it: education operates on our intuitions *ex ante* and *ex post*. The *objects* of education are the content of particular moral judgments as well as the reasons that bear on them, and this whole process can be *monological* as well as *dialogical* in form.

None of the things just said are meant to deny the shortcomings or detrimental effects explicit reasoning can have. The phenomenon of "choice blindness" (Johansson et al. 2005) illustrates just how poorly people sometimes reason about their decisions. Timothy Wilson and his colleagues (1989) found that the inability to accurately introspect the factors that influence our choices can lead to cases of severe attitude/behavior dissonances: the reasons people can articulate to themselves can influence their *assessment* of states of affairs, but leave their *behavior* almost entirely unaffected. It gets even worse when conscious reasoning not only does not adequately represent what influences subjects' beliefs and decisions, but starts to actually reduce their quality.[7] Wilson and Schooler (1991) have found that people's assessments of the quality of different brands of jams became worse, compared to experts' opinions, when they were asked to analyze why they felt the way they did.

Perhaps most troublingly, we are all familiar with many cases in which moral reasoning clearly does change people's moral judgment, albeit *for the worse*. We frequently use our powers of reasoning to tell ourselves moral lullabies that allow us to spin the moral truth into something more conveniently self-serving. Moral reasoning can often be used to minimize the consequences of one's action, dehumanize others, or diffuse responsibility, thereby aiding moral disengagement (Hindriks 2015; Bandura et al. 1996) Conscious deliberation can sometimes be disruptive. However, I see little reason to think that this is the rule rather than an exception, especially since most of the evidence on people's reasoning falling short is generated under rather unusual experimental conditions.

As far as their *post hoc* nature is concerned, moral reasoning and ordinary reasoning are "companions in guilt" (Lillehammer 2007), and the SI model uses a double standard for the two. The suggestion made by the model is that moral reasoning is prone to confabulatory after-the-fact rationalizations, because our moral intuitions are rooted in intense emotions, and we are strongly motivated to hedge them as well as we can. But this cannot be the correct explanation. Many psychologists nowadays (Wilson 2002; Wegner 2002) claim that all reasoning, not just the moral kind, is a *post hoc* enterprise, even though those types of reflection typically do not serve to protect dearly held ethical convictions. Reasoning comes after the fact due to simple restrictions of information processing capacities. But humans have to struggle with these restrictions in moral and nonmoral cases alike. Unless one wants to defend the idea that there is no such thing as genuine reasoning at all, whether it is moral or nonmoral reasoning, the insight that all reasoning is epiphenomenal in this sense should not be particularly troubling. In fact, it is hard to see how it could be otherwise.

Conclusion

The primary function of explicit moral reflection isn't to directly precede and thereby cause people's moral judgments, but to feed back into people's intuitive responses and to improve, shape, and inform them. It is an ongoing process that creates a chain of feedback loops, with each one influencing the following. The reason why the SI model mistakes this process as confabulation is that if one looks at only one of those loops, it is indeed the case that the underlying intuitive process is prior to subjects' conscious reasoning: for each loop at a time, the automatic intuition comes first. But if one steps back and takes a look at the whole chain of feedback loops, what used to look like idle confabulation suddenly starts to look like an extremely efficient way of managing one's intuitions. This, I have argued, is what the education of our intuitions is all about.

3 Moral Reasoning as a Social Practice

Introduction

When asked about the origin of his moral values, Brian Palmer, a middle-aged business man from San Jose, California, had little to say: "Why is integrity important and lying bad? I don't know. It just is. It's just so basic. I don't want to be bothered with challenging that. It's part of me. I don't know where it came from, but it's very important" (Bellah et al. 1985, 7). But then again, why would he want to challenge his beliefs? Does he have a reason to do so? Which thought could convince him—or any of us—to question that lying is bad and integrity valuable?

Morality pervades our lives. It structures our everyday experiences as well as our social interactions, and we are faced with morally relevant decisions all the time. These decisions require us to make moral judgments about what is right or wrong, good or bad, or simply the thing to do. Some people may believe that we make the moral judgments we make because we have good reasons to do so, and that we arrive at our moral judgments on the basis of genuine reasoning and careful deliberation. But Brian Palmer's concession as well as the psychological findings I have cited in the previous two chapters suggest that it might not be so simple; in fact, it seems to be the other way around. We reason the way we do because we already happen to have certain moral convictions that are ultimately grounded in something that is itself groundless: our most basic moral intuitions. And when we justify these intuitions, we do not do so in an impartial pursuit of truth, but to persuade others to accept our values.

This chapter is about moral reasoning and its relation to subjects' educated moral intuitions. The anti-rationalist challenge has two parts: the first, negative part consists in a debunking of moral reasoning as a confabulatory *post hoc* enterprise. The second, positive part consists in an account of moral reasoning as something that is not about individual people seeking truth, but about socially embedded people seeking allies. I have argued that the first part of this challenge to rationalism does not succeed. In what follows, I will argue that the second does not succeed either. I show that the SI model is right in that moral reasoning is a social practice, but that it draws all the wrong conclusions from this. Where the SI model sees strategic persuasion, we should see cooperative reasoning.

The "social" part of the anti-rationalist challenge is based on mistaken assumptions about structure and practice of moral justification. I want to flesh out a sufficiently rich account of moral justification, and shall bring a number of normatively useful distinctions from the epistemological literature to bear on the empirical issues that interest me in this book. I shall refer to the account I intend to sketch as the *challenge and response* model of moral justification (see Brandom 1994; Williams 2001; Willaschek 2007; Thomas 2011). This account is supposed to answer two questions. The first one is: why do people rely on their educated moral intuitions in their moral reasoning? Why does their reasoning come *post hoc,* after their intuitions? The second is: how should we understand the social dimension of moral reasoning?

This chapter has nine sections. In section 1 I analyze which picture of moral justification stands behind the SI model's charge against rationalism. Sections 2 and 3 show that there is an alternative to this picture, and how this alternative ties in with the educated intuitions account. Section 4 raises more doubts about the social intuitionist interpretation of dumbfounding experiments. Sections 5 and 6 describe the practice of moral reasoning in more detail, while sections 7 and 8 explain why this practice has to be *social.* In section 9, I apply the account of moral reasoning developed in this chapter to the empirical psychology of moral judgment and reasoning.

The basic outlines of my approach are simple enough: we are typically justified in holding our moral judgments on the basis of automatic moral intuitions. Moral reasoning terminates with basic moral intuitions, which in a given context of inquiry do not stand in need of justification. What stands in need of justification depends on the availability of legitimate challenges. If there is no such challenge, no reasoning is required. That is why subjects can make moral judgments on the basis of their educated intuitions, rather than costly antecedent reasoning. We do not reason out of and into the blue; rather, the specific goals and standards of moral reasoning are determined by the particular challenge put forward against one's moral intuition. Finally, episodes of reasoning do not terminate with a fixed type of moral "regress stopper": what a chain of reasoning legitimately terminates with depends on the particular challenges that set off the chain. All of this taken together reflects the *challenge and response* structure of moral justification.[1] Challenges are put forward by a social audience; responses to those challenges are directed at this audience. It is this social structure of genuine moral reasoning the social intuitionist model of moral reasoning fails to take into account properly—at great theoretical cost.

Let me start with a warning. The interdisciplinary nature of this approach—located somewhere between moral psychology and moral epistemology—makes it inevitable for me to sidestep many important issues worthy of discussion. The account I wish to develop remains silent about many problems concerning the most important issues in moral epistemology: I will not be able to properly deal with the internalism/externalism debate, examine the questions that have to do with foundationalism, coherentism, and its rivals as rigorously as they need to be examined, or to delve into the intricate issue of moral skepticism. It is not my goal, nor could I ever hope to achieve it in this book, to develop a theory of moral knowledge. My goal is far more modest: I wish to bring certain distinctions made within moral epistemology to bear on the psychological issues I am primarily interested in, and show why it makes sense (from a philosophical point of view) for ordinary people to reason the way they do.

1 Moral Intuitions and the Structure of Moral Justification

The kernel of truth in the SI model is that moral judgments are typically made automatically. But this should not worry us: it would quite simply be way too cognitively demanding to arrive at one's moral assessments of situations via effortful online reflection. Conscious reasoning must remain the tip of the cognitive iceberg. I have suggested, however, that there is nothing wrong with this for a second reason: most of our higher cognitive functions are acquired, or at the very least shaped, over the course of our upbringing. The primary function of moral education is to equip us with habitualized patterns of moral judgment formation that allow us to navigate our everyday world without having to constantly embark upon costly deliberation. What once was cognitively effortful (System II processes) has become habitual and automatic over time (System I processes): we have no reason to think that in this migration process episodes of reasoning transform from rational to irrational. Moral judgments are made on the basis of educated moral intuitions. The *challenge and response* model of moral reasoning holds that these intuitions are presumptively justified: typically, the entitlement to accept them does not have to be earned through explicit *before the fact* reasoning. In this section, I wish to elaborate on these claims, and explain how they contradict some of the main assumptions behind the anti-rationalist challenge.

Let us return to the SI model for a moment. Remember that Haidt's main claim is that subjects' moral judgments are not based on reasoning. He concludes that this is the case because his "dumbfounding" experiments seem to show that people form an automatic, intuitive judgmental response to a morally salient scenario, and that they engage in moral reasoning only *post hoc*. It is not like subjects do not reason at all—they *do* justify their judgments, albeit after the fact. Yet Haidt takes this as evidence that subjects' judgments are not based on reasoning. For their judgments to be based on reasoning, according to Haidt, one must presumably engage in such reasoning *ante hoc*. Justification, for the SI model, requires *before the fact* reasoning.

Moreover, Haidt says that people do not reason in the pursuit of truth, but because they want to defend the emotionally charged intuition they already happen to have. Since this does not count as a genuine truth-seeking endeavor, he seems to suggest that the *post hoc* method people follow when they reason about their moral beliefs is "unreliable" in some sense. It is not guided by normative principles, but by emotional interests. So according to the SI model, genuine reasoning is always *before the fact* reasoning, and reliable judgment formation requires this form of reasoning. This is the idea I want to develop an alternative to.

What does it take for a judgment to be justified? In order to state this question more precisely, let us first distinguish between *personal* and *evidential* justification (Williams 1999, 187–188). Personal justification is about cognitive responsibility. A person is justified in believing something when she did not behave epistemically irresponsibly in acquiring the belief.

Evidential justification, on the other hand, is about adequate grounding: one is evidentially justified in believing something if one has grounds that make it likely to be true. How do the two forms of justification relate to each other? According to Michael Williams's diagnosis, many accounts of knowledge establish a very tight link between the two, arguing that the former requires the latter. Williams calls this link the "prior grounding requirement," according to which justification requires the "possession of reasons" (2001, 148).

Now this is just a different formulation of the requirement that, for a subject to have justification for a judgment, she must have earned justification through explicit reasoning from consciously accessible reasons. In other words: for a judgment to be justified, one must have reflective access to the set of reasons that justify the judgment. On the other hand, that moral intuitions enjoy *presumptive* justification means that this justification does not have to be earned through explicit justification, or, as the prior grounding requirement has it, "possession of reasons." Note that the *challenge and response* model I will outline does not deny that entitlements accrue to subjects on the basis of cognitively responsible behavior. It merely denies that cognitive responsibility is exclusively tied to conscious reasoning from internally accessible considerations.

The anti-rationalist challenge rests on the heavy-duty assumption that entitlements to (moral) judgments *always* have to be *earned*. The idea is that justification requires a series of *positive steps* which subjects have to undertake in order to be entitled to a particular judgment. According to Williams, "the difference between the 'Prior Grounding' and 'Default and Challenge' conceptions of justification is like that between legal systems that treat the accused as guilty unless proved innocent and those that do the opposite, granting presumptive innocence and throwing the burden of proof onto the accuser" (Williams 2001, 148). One can see that the idea that justification always requires one to actively justify one's judgments is a commitment that is deeply entrenched in philosophical and psychological thinking. And it seems plausible to assume that this commitment fuels the social intuitionist idea that if subjects do not actively reason toward to their moral judgments *before the fact*, their judgments are not justified. The possibility that to be justified might be the rule, rather than the exception, and that explicit moral reasoning is only required under special circumstances, is never even considered.

The *challenge and response* model is usually seen as a response to general philosophical skepticism. My claim is not that the SI model is a form of moral skepticism, nor am I interested in offering a response to moral skepticism. I will leave this issue aside here. What I am interested in is to excavate some of the structural similarities between notions of reasoning that lead to skepticism and the social intuitionist model. Moral skepticism and the main elements of the SI model are both based on and motivated by the same intellectualist account of justification. This account imposes overly demanding standards upon human behavior, standards that we are given no independent reason to accept over and above a philosophical tradition that values self-reflection over habit, and conscious reasoning over uninhibited action. It involves, among other things, a commitment to Williams's "prior grounding requirement," which is translated into the psychological idiom in terms of what I have called AR and CR. To circumvent these intellectualist pitfalls, one needs to look for the best available response to skepticism in general, see whether it avoids the internalist assumptions that lead to (moral) skepticism, and hope that this response will also allow us to say something meaningful about the psychology of moral reasoning.

2 Structural Contextualism

The purpose of this section is to explain what the notion of justificatorily basic moral intuitions contributes to our understanding of the structure of moral reasoning. Many philosophers have shared the sentiment that the demands made by moral skepticism are overly stringent. In fact, this stringency is *the very point of skepticism*. One way of answering the skeptical challenge is by meeting it straightforwardly, by arguing that we can meet the required standards of justification. That is what epistemological foundationalists, coherentists, and infinitists attempt to do (Bonjour 1976; Davidson 1989). A second way would be to argue that a particular set of standards raises the bar too high. This different and, in my view, more promising strategy is to deny that the proposed standards are valid in the first place.

Skeptical scenarios (think of Descartes's "evil demon" or Putnam's "brain in a vat") are the most natural way to generate the skeptical challenge toward *perceptual* beliefs, and it is difficult to see how one could undermine the evidential basis of one's *moral* views by pointing out that one might be fed fake sensory perceptions about the external world. (Unless, of course, these illusory perceptions undermine the factual basis of one's moral beliefs.) Structural skepticism, on the other hand, generates the skeptical challenge from the regress problem, and this problem is indeed very relevant to accounts of moral reasoning. Consider this example: you find yourself in a concrete situation that calls for a moral verdict. You make one. You think that, say, it is wrong for the top 1 percent of the people in a nation to possess 50 percent of that nation's wealth. But here is someone who challenges your judgment, and asks you for a reason. You respond by saying that you think this distribution of money and resources is wrong because it is unfair. Well, why is it unfair? How do you know? And so off the regress goes. This is where the contextualist account of moral justification comes into play. Its main goal is to offer a contextualist response to the problem of potential regresses of justification.

Moral reasoning can go in circles, go on infinitely, or stop at an arbitrary point. None of these options seem particularly promising. But why start to trace one's judgments back to their justificatory foundations in the first place? Here is Sinnott-Armstrong's simple and powerful reconstruction of the argument that aims to show why we are forced into this vicious regress and where it ends:

(1) If any person S is justified in believing any moral claim that p, then S must be justified either inferentially or noninferentially.
(2) No person S is ever noninferentially justified in believing any moral claim that p.
(3) If any person S is justified in believing any moral claim that p, then S must be justified inferentially.
(4) If any person S is inferentially justified in believing any moral claim that p, then S must be justified either by an inference with some moral premises or by an inference without any moral premises.
(5) No person is ever justified in believing any moral claim that p by an inference without any moral premises.
(6) If any person S is to be justified in believing any moral claim that p, then S must be justified by an inference with some moral premises.
(7) No person S is ever justified in believing a moral claim that p by an inference with a moral premise unless S is also justified in believing that moral premise itself.
(8) If any person S is justified in believing any moral claim that p, then S must be justified by a chain of inferences that either goes on infinitely or includes p itself as an essential premise.

(9) No person S is ever justified in believing any moral claim that p by an inference that includes p itself as an essential premise.
(10) No person S is ever justified in believing any moral claim that p by a chain of inferences that goes on infinitely.
(11) No person is ever justified in believing any moral claim. (Sinnott-Armstrong 1996, 9)

The *challenge and response* model denies certain readings of (2), (3), and (7). Propositions (2) and (3) can be seen as restatements of the accessibility and the causality requirement or, in Michael Williams's idiom, the "prior grounding requirement." Evidently, they deny that there are any justificatorily basic (moral) judgments. Proposition (7) is based on the assumption that it is always, that is, even in the absence of positive reasons to do so, legitimate to ask for a justification for someone's (moral) beliefs—a thesis that nicely complements the second and third premise in the above argument. The *challenge and response* model denies all of this, as it holds that there can be beliefs which are not in need of justification, yet can convey justification on other, nonbasic beliefs. Within our actual practices of moral reasoning, there is no requirement to respond to "naked challenges" (Williams 2001)—error possibilities we are given no actual, *positive* reason to believe obtain.

It is worth pointing out that the most elaborate account of moral contextualism, Mark Timmons's (1996, 1999) *structural contextualism*, shares the commitment to normative *and* descriptive adequacy adopted here. In describing the building blocks of his contextualist epistemology, he explicitly cites the phenomenon of "moral dumbfounding" *avant la lettre* as evidence for the fact that human moral agents reason from basic moral beliefs which they take not to stand in need of justification in a given context. It is interesting to note that Timmons describes the phenomenon in almost the exact same way as Haidt does, but draws entirely different conclusions from it. Following Brian Palmer's words I have quoted above, Timmons writes:

> My reading of all this is ... that, in many contexts at least, there are moral beliefs—general moral beliefs—that provide the basis for one's coming to justifiably hold other moral beliefs, but beliefs for which most ordinary people have no (justifying) reason. In these (rather typical) cases, I do not think it is plausible to criticize such agents for being epistemically irresponsible. They have particular moral beliefs that rest for their justification on other moral beliefs that represent the core of their moral sensibility. (Timmons 1999, 216)

We can see from this quote that a lot hinges on the account of justification invested in one's reading of the data. Timmons and Haidt do not seem to disagree about the facts: at a certain point in their moral reasoning, people hit rock bottom. Timmons and Haidt merely disagree about what it takes for subjects to justifiably hold their moral judgments.

It seems that it can never be epistemically responsible to make unsupported judgments or to hold beliefs one does not have any justification for. But, the contextualist teaches us, this is an illusion we have only become convinced of by tacitly subscribing to certain false assumptions about what makes for sufficient justification. In fact, we become initiated into an intricate array of practices of moral and nonmoral judgment as a result of our upbringing, and we take these practices as a starting point that we must strive to improve upon from within. Questioning the whole practice all at once gets us nowhere. We need to treat the basic beliefs that determine the rules and the content of those practices as basic in a justificatory sense, and as conveying justification on other, less basic beliefs. This does not mean that the basis of our reasoning lies outside the game of giving and asking for reasons in a strict and ultimate sense. Given a legitimate, positive reason to think that certain contextually basic beliefs ought to be questioned, the focus of inquiry can shift onto them. However, a simple, unsupported "Well, how do you know?" will not do that trick.

Timmons argues that the architecture of moral reasoning is built up out of contexts that comprise epistemically dependent beliefs—which stand in need of justification—and epistemically independent beliefs—which do not. His thesis can be decomposed into the following three claims (see Timmons 1999):

(1) Epistemic responsibility does not (always) require justifying reasons.
(2) Some beliefs which are responsibly held without any justifying evidence can justify holding other beliefs.
(3) Which beliefs play which role depends on contextual features.

Propositions (1) and (2) stop the regress of justification. In a nutshell, they say that one can responsibly make certain judgments either in the absence of any reasons whatsoever, or at least on the basis of reasoning chains that eventually come to an end. Think about the short conversation from above. According to (1)–(3), the reason "because it is unfair," does justify the person's moral judgment about the situation, provided that it does not stand in need of justification itself. If it doesn't, it is contextually basic. The mistake the skeptic makes is to assume that all beliefs always stand in need of justification.

3 Moral Justification and Moral Education

How does the idea that moral judgments are based on moral intuitions, which subjects are usually entitled to, tie in with the idea that these intuitions are subject to education? The concept of education has a twofold sense. First, it refers to education as such, the process in the course of which one is first introduced into a certain set of practices—the process of becoming a morally judging subject in the first place. Second, it refers to the ongoing improvement of one's moral intuitions through the feedback loops that continuously shape them, which happens once one has already reached the stage of full-blown moral agency. The fact that entitlements to judgments simply accrue to us does not mean that one does not have to satisfy certain criteria of competence in order for this to be so. Williams and Timmons explicitly mention the importance of education for their account of justification and entitlement. Williams's slogan for this is: "The status of an epistemic subject does not come with mere sentience: it has to be earned through training and education" (2001, 149). That is to say that being a subject that is capable of having full blown moral intuitions and making provisionally justified judgments is not something that is entirely disconnected from inferential justification. It first requires an initiation into the "space of reasons" that is constituted by shared conceptual practices, the ability to competently navigate this space and a minimal grasp of the inferential structure of the concepts employed in one's judgments.

One has reached the point of competence as a moral judge when one has acquired what Timmons calls a "moral outlook." Such an outlook is constituted by a set of educated moral intuitions, judgmental habits, and associated emotional responses that determine the justificatory horizon from within which people justify their particular judgments once they have been called into question: "A moral outlook represents a way of viewing and responding to one's environment from a moral point of view. ... One comes to have a moral outlook through a process of moral education" (Timmons 1996, 311). According to Timmons, this process includes at least four different elements. Acquiring a moral outlook consists in (see Timmons 1996, 311):

(1) the acquisition of a sensitivity to the morally relevant features of situations;
(2) learning to associate appropriate emotional responses with the objects of one's moral judgments;
(3) the acquisition of a set moral generalizations that specify what the morally relevant features of a situation are, as well as being acquainted with so-called moral exemplars, paradigmatic cases of (im)moral actions or traits; and, finally,
(4) the acquisition of basic skills of moral reasoning, like coming to understand and be able to adopt an impartial perspective, universalization/reversibility reasoning and moral *judgment* (the skillful application of general moral principles to particular cases).

The habitualized moral responses one acquires over the course of one's moral education then function as the "hinge propositions" that organize one's moral judgment and reasoning.

4 Confabulation or Inarticulateness?

As the saying goes: Never attribute to malice that which can be adequately explained by stupidity. I wish to suggest, on the basis of the educated intuitions account of moral judgment, never to attribute to confabulation that which can be adequately explained by mere inarticulateness.

The educated intuitions account holds that moral reasoning figures in the acquisition, formation, maintenance, and correction of people's moral intuitions. These patterns of reasoning are habitualized over time, and are being actualized in particular instances of judgment formation without drawing on scarce cognitive resources, and by bypassing effortful conscious information processing. These reasons are nonetheless causally effective when it comes to what judgments subjects endorse. The SI model overlooks the possibility of a migration of explicit reasoning into our automatic responses, and instead suggests that typically, subjects confabulate and produce sham after-the-fact rationalizations for their irrational gut feelings.

But note that all of the evidence there is for the claim that in *typical* cases, subjects confabulate, is based on data gathered in the context of situations that are rather *atypical*. This is because confabulation is defined as behavior in which subjects cite considerations as relevant for their judgment and conduct that, unbeknownst to them (but known to "us" scientific observers), are clearly causally disconnected from people's behavior, because the cited factors do not in fact obtain in that situation. I suspect that it would be impossible, for methodological reasons, to provide direct evidence for the idea that subjects are typically confabulating in ordinary cases in which they cite considerations pertaining to factors that *do* obtain, and in which their responses and their reasoning are not so disconnected. What reason would there be to think that in such typical cases, subjects are making anything up?

Consider the following examples by Terry Horgan and Mark Timmons (2007): first, imagine a woman who calls her longtime friend to tell her about a promotion she just received. As her friend picks up the phone, she immediately notices the sad tone in her voice. She asks her what the problem is, and hearing that she has just been fired from her job, the woman immediately decides that this is not the right time to brag about her professional success. Later, she thinks that she did not bring up the topic she originally wanted to talk about because it would not have been the right time, she did not want to hurt her upset friend, and because it would have been callous and insensitive to mention it. Second, imagine a man who is climbing a mountain when he hears a woman screaming. As he comes closer, he sees that the woman is being attacked by a male aggressor whose intentions are obviously dubious. He spontaneously decides to help, hoping that his intervention will make the man go away. Later, he thinks that he acted the way he did because the woman was in dire straits, he was the only one around, and he did not want to be a coward. In each of these cases, the reasons the woman and the man cite for their actions *after the fact* did not figure in their conscious deliberation that led them toward acting the way they did. In fact, they did not consciously reflect about what to do at all. But it would be far-fetched to suppose that they were confabulating (in any meaningful sense), because their reasoning cites features of the situations that are both morally relevant and were, in all likelihood, causally responsible for their behavior. If, in everyday cases like these, subjects arrive at their judgments intuitively and automatically, and adduce their reasons *ex post*, and if there is no reason to think that these subjects are confabulating, then we need some account of how moral reasoning can influence moral judgment beyond conscious deliberation. The best explanation for this seems to be that people's reasons have shaped their automatic responses beforehand. Their reasons have *educated* their intuitions.

Now let us take another look at some of the artificial scenarios philosophers and psychologists have designed in order to accuse subjects of moral confabulation.

Trolley and *Footbridge*: In *Trolley* as well as in *Footbridge*, five lives versus one are at stake. In the first scenario, subjects are asked to judge whether it would be morally appropriate (the scenario used in the experiments does not distinguish between appropriateness as permissibility and appropriateness as obligation, which is what maximizing consequentialism would prescribe) to prevent a runaway trolley from killing five workers by diverting it to a side track, thereby killing one person on the side track. In the second scenario, subjects are asked to assess the moral status of pushing a very fat man onto the tracks, thereby sacrificing the man, but saving the five workers. Ordinary people as well as many trained philosophers have different intuitions about the *Footbridge* scenario than they have about the *Trolley* scenario. Most people think that it is permissible to divert the trolley, but impermissible to shove the man down the bridge, the identical body count notwithstanding. Some philosophers have argued that this is perfectly reasonable, as the latter case violates the principle of double effect, which says that harm brought about as a foreseen but unintended side effect can be permissible, whereas intended harm cannot.

Greene (2008) has argued that this is a case of rationalization, because subjects' responses are best predicted not by a conscious application of the doctrine of double effect to the case at hand, but by an emotional aversion toward pushing a person to his death, which they rationalize after the fact. In reality, he argues, there are no morally relevant differences between the two scenarios, and therefore, most people's nonconsequentialist intuitions are unjustified. Subjects' "deontological" reasoning is confabulated, because it does not cite considerations that, as the empirical evidence suggests, played a causally effective role in the formation of their judgments.

Incest and *Cannibalism*: The first of these two vignettes describes a pair of siblings who decide to have sex with each other while on vacation. They use birth control, do not tell anybody about it, enjoy it, and continue to have a healthy relationship with each other. The second one describes a vegetarian who works in a medical research lab when one day, she notices what a waste it would be to throw away the human cadavers she works with, takes a piece of the human flesh home with her, cooks it, and eats it. Nobody finds out about this, and she thoroughly cooks the meat so she does not get sick. Haidt and his colleagues describe these actions as victimless crimes, scenarios that were "carefully constructed" not to be harmful in any way.

One thing that tends to be dismissed rather than overlooked in the interpretation of this research is the phenomenon of cognitive resistance (Kennett 2011). People are asked to judge the described actions solely on the basis of the information they are given in the story. But of course, this is psychologically difficult, if not impossible. People rapidly and automatically fill in a lot of what they think is plausible. They are asked to buy into a story that describes two siblings who have a perfectly healthy relationship, *yet decide to have sex because it would be fun*. We can safely assume that subjects cognitively resist this description, because they regard it, plainly and simply, as impossible. People are asked to imagine a person with a character that allowed her to become a medical professional and the moral integrity to be a vegetarian, *yet decides to eat the flesh of a human cadaver*. To arrive at real moral judgments, subjects must be able to buy into the story as it is presented to them. They must take the characters for real people wrapped up in real situations. And for real people to do such things there *must* be something else going on with them. This "something else" is of course hidden, and since in their explicit responses, people are asked to rely only on the information they are actually given, these hidden reasons cannot be articulated by them. But that does not mean that they are not there.

Famously, Greene has argued that one can "spot a rationalizer without picking apart the rationalizer's reasoning" (Greene 2008, 67). But what if the alleged rationalizer's reasoning is in fact perfectly valid? What evidence do we have left for the claim that the subject is a rationalizer, rather than just a reasoning subject? The claim that people irrationally cling to their moral feelings and produce superficially convincing, but easy to debunk *post hoc* rationalizations is based on the observation that people justify their judgments *in the absence* of good reasons to do so. In all of the above cases, Haidt and Greene argue that people mistakenly condemn "harmless" actions (or actions that would cause significantly less harm than an alternative, respectively). There are two main problems with this claim (in what follows, I will draw heavily on Daniel Jacobson's [2012] analysis of moral dumbfounding): first, it is suggested that

(1) the only good moral reasons there are, are harm-based reasons.

This point is especially important for people who subscribe to a deontological theory in normative ethics. For the sake of the argument, if we grant that deontology is the theory that holds that some act types are morally wrong *regardless* of their good or bad, harmful or harmless consequences, then it cannot be argued without begging the question that deontological theories are unjustified because they license the condemnation of actions that do not have harmful consequences. That is the very point of deontological ethics.

Second, and more importantly, Haidt's and Greene's arguments rely on a conception of harm too narrow to accommodate distinctions which virtually all available moral theories consider to be relevant. There are distinctions between

(2) (a) actual and expected harm,
(b) harm that does not pertain to subjective disutility,
(c) indirect forms of harm, and
(d) actions that are harmless but less beneficial than an available alternative.

Third, there seems to be no room in Haidt's social intuitionist or Greene's dual process model for the idea that,

(3) a person's traits and motives might matter to the moral evaluation of her actions and character.

Clearly, Julie and Mark's actions in the *Incest* scenario are very risky (as captured by 2a), they *could* easily result in emotional damage that far outweighs the pleasure they might have caused at the time (as captured by 2b), they are prohibited by the best set of rules concerning what ought and ought not to be done (as captured by 2c) and they are most certainly less beneficial than it would be to simply not do it (as captured by 2d).

Now take a look at some of the examples that populate the psychology of moral judgment: clearly, that one ought to be readily prepared to kill a person in order to achieve a certain desired effect is not very good advice, it is not the best policy one might adopt and it certainly violates that person's rights, which might not be a conclusive reason, but is hardly an *irrelevant* consideration. Clearly, it would be frivolously careless to risk the incomparably important relationships one has to one's siblings simply because it would be exciting and fun to try something new, even if things turn out to be fine. Clearly, it would not be virtuous to wantonly desecrate an item (such as a flag) that stands for something one cares deeply about, and thereby show that one is willing to symbolically express contempt for something significant in exchange for a small amount of money.[2] And clearly, if you have no problem whatsoever with pushing a man to his death, then something is seriously wrong with you *even if* it might in fact be the best thing to do in the circumstances.

Some of the above distinctions are more sophisticated, some less, but it would be far too demanding to require of an ordinary experimental subject that she come up with a justification in the above terms on the spot. Given that the above scenarios are extremely far-fetched and unusual, and given that there still are many good reasons that can be found for condemning the proposed actions, it does not seem like the best explanation to attribute subjects' behavior to genuine confabulation, rather than mere inarticulateness. Obviously, subjects have very strong intuitions that the described actions are morally dubious, and try to justify these intuitions after the fact. But this is not because they have some brutish disgust response toward incest which they try to defend no matter what, but rather because they have acquired an intuitive sensibility toward the morally relevant features of a situation over the course of their upbringing, the inner workings of which often are not readily accessible to them.

5 From Challenges to Responses

Without our educated moral intuitions, we wouldn't have anything to start from in trying to figure out what the morally right thing to do is. No moral code can be designed from scratch, and no moral code can be turned into an object of our critical practices except from within.

So far, I have said various things about the structure of justification, its rationale and its function. But all these remarks have remained fairly general. They apply to justification as such, regardless of whether we are dealing with the moral or nonmoral domain. And this is a perfectly legitimate strategy: that justification starts from provisionally accepted intuitively compelling judgments is not a feature that is specific to the moral realm. In this respect, moral reasoning and all other types of justification are "companions in guilt" (Lillehammer 2007).

Timmons is optimistic about the fact that we can offer a nondeflationary solution to the *regress problem*: we can stop the regress of justification by arriving at beliefs that do not, on a given occasion, stand in need of justification. But according to Timmons, the regress problem is still a *real problem*. Advocates of the *challenge and response* model disagree with this. Here is how Robert Brandom puts it: "One of the lessons we have learned from thinking about hyperbolic Cartesian doubt is that doubts too sometimes need to be justified in order to have the standing to impugn entitlement to doxastic commitment. Which commitments stand in need of justification is itself a matter of social practice" (Brandom 1994, 177). There are moral beliefs to which subjects are *prima facie* entitled. It is misleading to suppose that it is always acceptable for someone to demand justification, although by good fortune, we do often arrive at suitable regress stoppers. Rather, we do not even have to start the regress of reasoning in the first place, because absent legitimate challenges, we are not required to. Timmons's theory remains silent about the possibility of such "first round" *prima facie* entitlements. It does not explicitly account for the fact that asking for a justification is what requires some kind of positive reason to think that a belief needs to be justified. If no such justification can be given, the challenge is "naked." The idea that for moral judgments, or any kind of judgment for that matter, to be justified, one's holding them must be the upshot of an explicit chain of reasoning undertaken in advance suggests that there is no reservoir of moral beliefs whose authority we can take for granted, and to which we can treat ourselves as being *prima facie* entitled. But this suggestion is problematic,[3] because it imposes a cognitive load too heavy for human moral reasoners to bear.

In this section, I want to argue that we ought to revisit the concept of a moral intuition from a *challenge and response* perspective. Psychologically speaking, intuitions are simply judgments we arrive at very quickly and effortlessly, and which have a certain initial compelling quality to them. Moral intuitions, as they have been traditionally conceived, have been said to be one or more of the following things:

(1) self-evident,
(2) self-justifying,
(3) indubitable, or
(4) the output of a special mental faculty.

Modern intuitionists typically avoid committing to any of these characteristics. Instead, they think about ethical intuitions as evaluative "intellectual appearances" (Huemer 2005, 101ff.) which are arrived at non-inferentially and are thus held *directly*, *firmly*, *pretheoretically*, and grasped with at least at *minimal understanding* of their propositional content (for these four criteria, see Audi 2005). I will not offer a comprehensive discussion of even the basic tenets of classical ethical intuitionism here. Modern intuitionism seems to avoid many of the problems classical intuitionism suffered from. But it remains true that modern intuitionists typically cannot be bothered to provide a positive psychological account of how intuitive moral cognition works and what role intuitions play in the game of giving and asking for moral reasons.

Our educated moral intuitions, as I understand them, need not have any of the above properties. Rather, the *challenge and response* account says something entirely different about intuitions and their place in the space of reasons: there are certain moral judgments which can be used as starting points for our reasoning, and cannot be challenged unless certain conditions obtain; there is a distinction between legitimate challenges which give one an actual positive reason to doubt a moral judgment, and naked challenges which just point to a logically conceivable error possibility for which there is no reason to think it might actually obtain. Moral intuitions need to be accounted for in terms of their inferential role within the game of giving and asking for reasons, rather than the faculty they stem from. They are nothing more than automatically made, *prima facie* compelling yet defeasible judgmental responses to morally salient situations.

6 From Responses to Challenges

The obligation to cite appropriate reasons in support of one's judgment is the kernel of truth behind internalist conceptions of justification and reasoning. In ordinary cases, subjects do not have to undertake cognitively demanding positive steps in order to justifiably hold their judgments. But not all situations are of the ordinary kind. In the light of relevant challenges, presumptive justification vanishes, and must be regained by adducing good reasons for one's beliefs. If no such reasons can be given, one can no longer justifiably hold the judgment in question.

What types of undermining reasons are there? And how does one go about responding to them? Here, we can draw on some suggestions made by John Pollock (1986, 1987):[4] according to his proposal, there are rebutting and undercutting defeaters.[5] (For reasons of convenience, I will refer to *defeaters* simply as *challenges*.) Rebutting challenges directly call into question the judgment at issue, either by providing evidence that the judgment is incorrect or by demonstrating that some other judgment—whose truth entails the falsity of the other one—is correct. Undercutting challenges, on the other hand, challenge the justificatory foundation of a judgment: without indicating directly that the judgment at issue might be false, they undermine the evidence there is for believing it. Suppose I believe that it is worse to hurt a newborn than a toddler; a rebutting challenge to that belief will point out to me that this is not the case, and provide some reason for the belief that it is in fact equally wrong to hurt a toddler. An undercutting challenge could consist in an argument that says that my belief is due to certain factors which are entirely morally irrelevant, such as the fact that I find newborns overwhelmingly cute, and toddlers utterly annoying. This undercuts the justification I have for my belief.

Now Pollock holds, and I shall follow him in doing so, that justification is a function of the availability of legitimate challenges. Obviously, a lot hinges on what exactly it is that renders a challenge legitimate. So far, I have argued that no "naked" challenge can ever be legitimate: the moral intuitions one has under epistemically favorable conditions cannot seriously be called into question simply by asking for some justification, without any reason to suppose that such a justification is required. It takes some positive reason to think that it is.

The dependence of justification on the presence of legitimate challenges can be recursively applied to the judgments originally challenged, the challenges itself and the responses to those challenges. Symmetrically to the distinction between rebutting and undercutting challenges, there are *challenge-overcoming* and *challenge-dissolving* responses (cf. Cassam 2007). To overcome a challenge is to meet it; to dissolve a challenge is to show that it did not pose a serious challenge in the first place. Due to the recursive structure of defeasible justification, the following three rules concerning the justificatory status of (moral) judgments apply to the practice of moral reasoning (cf. Pollock 1987, 7):

A (moral) intuition is justified if

(1) it is presumptively justified or supported by at least one unchallenged chain of reasoning.

A moral judgment is unjustified if

(2) it is not presumptively justified or a chain of reasoning contains a challenge to the judgment that has not been adequately responded to.

And what it means to regain justification for one's judgment is stated by:

(3) If a chain of reasoning contains no challenges that have not been adequately responded to, then the (challenged) judgment is (presumptively) justified.

In a nutshell, this means that a judgment is justified if and only if there are no legitimate undercutting or rebutting challenges to it, or if all challenges have been overcome or dissolved. If there are any undercutting or rebutting challenges to the judgment, and at least one of these has not been overcome or dissolved, then the judgment is unjustified. There are thus

(1) presumptively justified moral judgments,
(2) rebutting challenges,
(3) undercutting challenges,
(4) challenge-overcoming responses (which are rebutting challenges to rebutting or undercutting challenges), and
(5) challenge-dissolving responses (which are undercutting challenges to rebutting or undercutting challenges).

With these distinctions in hand, we can also make more sense of the justificatory "basicness" of certain moral beliefs Timmons and Williams are talking about. Pollock describes presumptively justified judgments as "initial nodes" in chains of inferences. On the assumption that it is correct that our practices of justification do not consist in giving reasons for each and every belief we hold, but only those which are legitimately called into question, it immediately follows that those beliefs which have not yet been called into question must be taken to be justified—for the time being. It is superfluous to postulate a special mental faculty to explain what a moral intuition is, and unnecessary to demand that moral intuitions be self-justifying or self-evident. Moral intuitions can be understood solely in terms of their position in the intricate architecture of the space of reasons.

7 The Flexibility of the Space of Reasons

The architecture of this space and the structure of moral justification is not only intricate, but also elusive, and in constant flux. The account of justification we have inherited from the philosophical tradition has it that there are different *types* of judgments, some of which—by their very nature—enjoy the privilege to be justifying, others to be the ones justified. This account agrees with the one developed above that there are basic and non-basic judgments, whose distribution in the web of our beliefs determines its overall structure. But it adds that where in that structure our beliefs belong is not determined by our justificatory practices, but by certain abstract features of the content of those beliefs. Thus perceptual beliefs about one's immediate surroundings are taken to be more basic than, say, scientific beliefs, and beliefs about sense data are said to be more basic than beliefs about ordinary facts of the external world.[6] In the moral domain, it is often thought that there is an analogous hierarchy ranging from particular moral judgments (*He ought to have done what he promised to do*) to mid-level moral principles (*Promises ought to be kept*) and, finally, abstract principle-generating rational procedures (*Keeping promises maximizes the overall welfare* or *The maxim to break promises under such and such circumstances cannot be universalized without contradiction*). But if the *challenge and response* model is correct, then this picture is misleading.

What a justificatory chain terminates with depends on features of the social practice of moral reasoning, such as what is at stake, at issue, or what particular question one is interested in. To use one of Wittgenstein's examples: if one is interested in checking whether one's eyes function properly, the fact that one can see that one has two hands justifies the belief that one's eyes do indeed work just fine. But if one wants to know whether it is raining outside, and tries to find out about the weather by looking out of the window, the fact that one's eyes function properly is taken for granted, and cannot be called into question by the observation that it is or is not raining. The structure of justification, that is, which beliefs function as basic, justification-conveying beliefs and which beliefs are being justified, depends on which open questions there are, and what the purpose of one's inquiry is. The same holds for moral justification. The assumption that there is only one type of moral reasoning—the one that traces back particular verdicts to mid-level generalizations, and these to their procedural foundations—does not do justice to the complexity of our justificatory practices.

Moral reasoning is thus challenge-dependent in a twofold sense: first, because explicit reasoning in support of one's educated intuitions is not required unless there is some positive reason to doubt them. Typically, the entitlement to hold a certain judgment need not be earned by fulfilling any special justificatory duties. Second, moral reasoning is challenge-dependent because subjects would not be in a position to know which direction to go in with their reasoning, if that direction were not determined by the specific content of the positive reason that was given in order to start off the reasoning. What elements a justificatory chain is required to contain, and what it legitimately terminates with, is specified by the challenge itself. Suppose subject A judges, on the basis of an intuitive response pattern she has acquired over the course of her moral education, that it was wrong for subject B to perform act X. Now there are several possible challenges available to a third subject C, who is doubtful of A's assessment of the situation. C could point out that the act performed by B was not, in fact, of type X, but instead of type Y, and argue that in the given situation, an act of type Y was not at all wrong. Or C could agree that the act performed by B was of type X, but mention some feature that renders X the morally appropriate thing to do under the circumstances. Or C could agree that the act was of type X, and that there is no such additional situational feature present, but cite some feature of B that shows why B was not responsible for the deed. (These three would constitute rebutting challenges.) Or C could hold that while A made the judgment in question, she was placed behind a filthy desk, surrounded by a terrible stench created by commercially available fart spray, and hypnotized into being extremely harsh and judgmental about other people's actions so that there is reason for her to reconsider her verdict (this would constitute an undercutting challenge). Anyway: each of the four challenges (and many more examples could be given) require a different kind of response to dissolve or overcome them. The question of what response counts as genuinely justificatorily basic in these four contexts, and what the structure of moral justification is as a whole, cannot be answered without taking into consideration the challenge-dependence of our reason-giving practice.

This gives us one more reason to think that the requirement to reason *ante hoc* when forming a moral judgment makes no sense at all. The SI model suggests that the proper way to arrive at a moral verdict is to reason from certain legitimate moral considerations—if we think about Haidt's *Incest* vignette (Haidt 2001), these include *it is impermissible to cause harm, the value of the family ought to be protected,* and so forth—combined with an assessment of whether these considerations apply in the case at hand, to a particular moral judgment. But upon contemplating the permissibility of consensual incest, why would one reason that way, rather than another? Why would one come up with the explicit, consciously held thought that it is impermissible to cause harm in the first place? Why *this* particular chain of reasoning, instead of any of the others also available to subjects? Without relevant challenges, there seems to be nothing to reason about.

None of this is to say that to reason from particular moral judgments over mid-level principles to abstract constructive procedures is always pointless. The point is that it is merely one form of moral reasoning among many others, and that it enjoys no special privilege among the many ways in which one can go about justifying one's moral intuitions. Sometimes it does become necessary to adopt a thoroughly reflective stance toward one's convictions: when this happens, and a major part of one's moral outlook is called into question as a whole, one must be able to see whether this particular part contradicts the highest principles and deepest foundations of morality one can think of. This is the scenario which the foundationalist picture of moral reasoning is tailored to account for, and where it has its proper place.

8 Giving and Asking for Reasons

Justifying one's moral judgments has to be understood in terms of what it is to justify them. The basic idea is that moral justification—being entitled to one's moral beliefs—is typically not earned by active online reflection. This does not mean, however, that reasoning does not play an effective role for one's moral beliefs: it does, either because it has played such a role in their acquisition, formation, maintenance, and correction or insofar as one's moral intuitions remain answerable to episodes of explicit reasoning.

But this form of reasoning is not required *ante hoc*, before we arrive at any moral verdict at all; rather, it has to do with being able to meet relevant objections (Annis 1978). If no such objection is or can be put forward, no reasoning is required.

Moral contextualism draws heavily on the idea of shifting standards: whether a subject is justified in some belief or not depends on how strict the epistemic standards are that apply in some context. Skeptical contexts invoke extremely strict standards which include all logically conceivable error possibilities, including the possibility that we are brains trapped in vats or in the pitfalls of our biological and cultural history. The structural contextualist replies that the standards applied by the skeptic do not improve our methods of inquiry by making them more rigorous; rather, they eliminate the hinges that need to stay put for the doors of our inquiries to turn.

But this does not mean that the structural contextualist—a version of which is the *challenge and response* model—does not have anything to say about shifting standards. In fact, the social flavor of the account allows us to incorporate this notion without its most damaging and counterintuitive consequences. Haidt's SI model defines moral reasoning out of existence: it assumes that, for our moral judgments to be based on reasoning, these judgments must be arrived at on the basis of conscious and, importantly, *monological* reflection. But, the SI model argues, both conditions are upset by reality. Moral reasoning is a confabulatory *post hoc* affair designed to persuade other people. That the concept of moral reasoning in terms of which the rationalist account has to be evaluated is monological is implicit in Haidt's definition of it: "*moral reasoning* can now be defined as a conscious mental activity that consists of transforming given information about people in order to reach a moral judgment. To say that moral reasoning is a conscious process means that the process is intentional, effortful, and controllable and that the reasoner is aware that it is going on" (Haidt 2001, 818). Using this mentalistic vocabulary makes no sense unless one understands moral reasoning as something that is going on within *one mind*.

The *challenge and response* model rejects the idea that for the rationalist about moral judgment, individual reflection is preferable to socially entertained reasoning. It agrees with the social element in Haidt's account, but attempts to describe it in ways that do not present a challenge to rationalism, but make it its trademark:

> Reasoning *always* has a social, argumentative function. ... The argumentative theory of reasoning posits that reasoning is a specialized mental mechanism that has evolved to find and evaluate reasons in argumentative contexts. Communication brings many benefits for humans, from gossip to increased ability to coordinate. Argumentation facilitates communication. Speakers can lay out arguments to convince skeptical audiences, and audiences can evaluate these arguments to decide whether to accept the conclusion or not. As a result, more propositions are successfully communicated and both the speaker and the audience are, on average, better off. But argumentation requires special skills, and reasoning is one of the mechanisms that evolved to allow humans to argue. (Mercier 2011, 135)

To make one's judgments and reasons explicit is to make them liable to social criticism, which then feeds back into our old moral intuitions and produces new ones with corresponding reasons. There is thus a deep connection between moral reasoning as a *post hoc* social practice and the education of moral intuitions.

To reason is to respond to salient objections. Who is in the business of raising objections to a subject's judgment? Typically, this will be done by other subjects. The *challenge and response* model of moral reasoning has it that relevant objections must express some sort of real and living doubt on the part of those subjects for their challenge to be justified. This real and living doubt—the positive reason to believe that the judgment in question is false or unjustified—determines the strictness of standards of justification as well as the structure of the response—the chain of reasoning it can legitimately terminate with. Take the following example by David Annis (1978, 215): I walk into a room with a friend, looking for a red chair, and I find one; my friend asks me how I know that it is really red; in order to be justified in my judgment that there is a red chair in front of me, I need not be in possession of superhuman perceptual capacities; I do not have to know about the physical properties of red light, or the psychological mechanisms by which I perceive it; and it is no matter of life and death I am dealing with. The pragmatics of what I am doing—that I am looking for something to sit on—specifies how strict the standards of justification are in this context and, accordingly, which error possibilities I must be able to rule out. But, as we have seen, the standards that apply to my endeavor are also partly determined by the objector group which holds me accountable. In this case, it is my friend. But during a physics examination or a medical procedure, the standards I will have to meet in order to achieve justification will be much higher (or at least different).

The structure of a response varies with the goals of the objector group as well. Suppose my friend, who is helping me with the search, points out to me that there is a red lamp shining onto the chair. In response to this challenge, neither excellent knowledge in physics nor freakish sensory abilities will do the trick. Instead, what might repair my justificatory status is if I went to the room yesterday, knowing that the light was out at the time, and still found the chair to be red. Both the structure and strictness of justification thus depend in many important ways on the social dynamics of the game of giving and asking for reasons: from the involved objector group to the social context in which the objection is raised.

Another reason to take the social dimension of moral reasoning seriously is the fact that we are just not that good at monological reasoning. Mercier (2011, 136ff.) points out at least two reasons why this is so:

(1) *Confirmation bias*. It is a notorious fact about human psychology that in evaluating the quality of their beliefs, subjects almost always look for confirming rather than falsifying evidence. This often has detrimental consequences. Mercier (2011, 137) mentions some of those consequences yielded by the confirmation bias, which are specific to monological reasoning: first, if only confirming evidence is revealed to subjects, they can become overconfident of their beliefs. Second, private reasoning can not only increase the degree of confidence in subjects' beliefs, but increase the severity of their content: as a result, subjects develop greater confidence in convictions that become more and more extreme, which in turn leads to a strong tendency to hold on to them even if they turn out to be false.

(2) *Satisficing*. People almost exclusively seek confirming evidence. And when it comes to the evidence they end up finding in that process, it is not the case that they are content only with the best possible evidence, but with reasons that are merely *good enough* to accept some belief.

Does socially undertaken reasoning work any better? I think it does. Take a look at confirmation bias: in social contexts, this vice becomes a virtue. At the very least, it becomes neutralized. The problem is canceled out through social interference effects: two individual confirmation biases do not add up to one big confirmation bias, but to a social microstructure that facilitates intersubjective criticism (Mercier and Sperber 2011, 65). Each subject tries to build the best possible case for *her* opinions. In doing so, both subjects inadvertently build the best possible case for *their* opinions.

What about the tendency toward satisficing? Satisficing arguments are very cheap. Sound and convincing counterarguments are cognitively costly. It is far easier to offload the obligation of their construction upon someone else; better yet to get by with a quick and mediocre argument. If Peirce is right, and the primary purpose of reasoning is to arrive at some belief that is in fact undoubted, then in episodes of private reflection, this satisficing method will often lead people to accept the first satisfying reason they can find. In social settings, however, this is not so easy: social criticism constantly confronts people's judgments with new information, thereby pushing them closer to an ideal state of truth and justification, rather than one of mere contentment.

(3) *Social extension.* I argued in chapter 2 that the shortcomings in people's moral reasoning that are dragged into the spotlight by the dumbfounding experiments should often be seen as a reflection of the fact that individual moral reasoning takes place within a social network of testimonially inherited justification. This picture suggests that moral reasoning literally takes place not within single minds, but *between* people.

In chapter 2, I have suggested a parity principle for automatic processes (PPA): mental operations that used to be consciously processed often migrate, over time, into our intuitive cognitive system, and there is empirical evidence that this process can be plotted as an asymptotic curve that relates increasing automaticity to the sheer number of times a process is being executed. The PPA suggested that if one used to be inclined to call the respective conscious process "rational," then the mere fact that a process has become habitualized should not undermine this inclination. The PPA thus helps defuse the automaticity part of the anti-rationalist challenge.

But if the picture of moral reasoning as a social practice is correct, then Clark and Chalmers's original parity principle can be used to help defuse the social part of the anti-rationalist challenge as well, because once we accept the possibility that moral reasoning takes place between people—and the aforementioned evidence suggests that it might well be rational to reason this way—we have no reason anymore to think that relying upon *other people's* expertise in moral reasoning somehow entails that *my* moral judgments are not based on reasoning. Andrew Sneddon (2011, 71–111) has recently integrated this hypothesis into his "wide moral systems" theory. He argues that the parity principle applies to social processes as well; moral reasoning does not have a primarily intersubjective function because it serves the purpose of social persuasion. Moral reasoning takes place within social structures, rather than isolated individuals who are trying to reach out to one another. And the parity principle for *social* processes (PPS) suggests that this fact alone gives us no reason to deny it any rational status, provided the processes at issue, were they to go on within one and the same mind, would not be denied this status. There is no reason to think that this type of cooperative reasoning threatens the main tenets of rationalism in any meaningful way. Rationalism is committed to the claim that moral judgments are based on the exercise of reason; it remains open to the possibility that empirical research will show that reasoning is socially exercised.

Some will inevitably think that my insistence on the advantages of cooperative reasoning paints an overly optimistic picture of the epistemic effects of social embeddedness. I will merely mention one particularly striking example to illustrate this worry: in his justly famous studies on peer pressure and conformity, Solomon Asch (1955, 1956) was able to show that a subject's social context can have very detrimental effects on the quality of her beliefs. In one experiment, a subject was placed in a room with a group of confederates; all of them were asked to judge which of two lines was longer, with the correct answer being blatantly obvious. However, when the confederates unanimously agreed on the *wrong* answer in one of the conditions, about a third of subjects went along with the obviously false (apparent) majority opinion. A recent study extends the Asch paradigm to moral judgment and found that conformity has a significant effect on people's (im)permissibility ratings of various standard moral vignettes as well (Kundu and Cummins 2008).

There are many other examples such as these. But even though they give us useful information about the strength of the effects of peer pressure and the conditions under which it is likely to exert the greatest influence, this provides us with no principled reason to be wary of the social structure of (moral) reasoning. The fact that social factors can often distort individuals' beliefs by no means entails that isolated, solipsistic reasoning is the better alternative.

Let me now respond to two other important objections. One often raised problem for the *challenge and response* model of (moral or epistemic) justification is the charge that the model merely reflects the pragmatic surface structure of justification, rather than what justification really is, and when it is "actually" required. This objection has been aptly put by Jarrett Leplin:

> What neither theory [Bayesianism and the default and challenge model] recognizes is that the contextuality of practice does not imply that initial epistemic status, in context, either lacks justification or requires none. That it is appropriate in practice *not to request* justification, or inappropriate to request it, does not imply that there need not be any, nor that there need be no theory of it. For even if justification required evidence, grounds, reasons, reliable sources of belief, foundations—whatever, it would *still* be necessary, in practice, to presume justification for lots of judgments and impracticable to insist that its basis always be delineated. A theory of when it is appropriate to challenge and defend beliefs is incomplete as a theory of justification, because there is more to justification that justificatory practice. Beliefs can be justified, and can need to be, even if it never becomes appropriate to challenge them, nor necessary to defend them. To grant beliefs a presumption of justificatory status no more obviates the question of what justifies them than a presumption of innocence obviates the question whether the accused is really innocent. (2009, 96)

That the *challenge and response* model merely reflects the surface structure of moral justification does not threaten the account: indeed, I take it that there is only that surface. The illusion that there is anything meaningful beneath that surface is created by adopting the "philosophical" perspective of someone who considers a case from the outside and grants himself full knowledge about all the relevant facts. In real life, however, we can never be in that position. In that sense, Leplin is wrong when he says that there is "more to justification than justificatory practice"—because there isn't—and that beliefs sometimes "need to be" justified even if it "never becomes appropriate to challenge them." In fact, this comes close to being incoherent. If they "need to be" justified, then there *must* be some appropriate challenge to them.

Another possible objection has it that if people can be presumptively justified in their moral judgments, this seems to allow them to hold on to those beliefs which are older, rather than better. I think this objection is mistaken as well. Remember that the *challenge and response* model says that subjects are entitled to hold unchallenged judgments. But a great deal of our ordinary moral convictions do not belong into that class. Huemer

(2005) identifies several challenges to common sense morality which are especially significant. There are challenges that pertain to

(1) the incoherence of (some of) our intuitions,
(2) cultural indoctrination,
(3) biological biases, and
(4) personal biases.

The pervasive incoherence of our everyday moral intuitions is illustrated, for example, by Peter Singer's (1972) observation that there are certain grave inconsistencies between the principles we endorse and the particular moral verdicts we make; in his now classic example, he points out that we treat cases in which we allow the death of a child which is close by and one which is on a different continent differently; yet *in abstracto*, we do not believe that spatial proximity is a morally relevant factor. The biases that come with a particular cultural background and the outright silliness of many of the values and practices we grow into, the morally reproachable tendencies of our psychology our evolutionary history has bestowed upon us, and the distortions we are liable to due to our egocentric and parochial perspective on the world are other examples for factors which we must be wary of when making moral judgments. But the point is that the *challenge and response* model does not condone an irresponsible reliance on whatever moral intuition one happens to have: one is only justified in believing something on the basis of those moral intuitions that have not been legitimately challenged or those that have survived this type of critical scrutiny.

The result will be anything but conservative, and will in all likelihood recommend serious revisions to people's common sense morality. Huemer discusses sexual morality as an example: people's moral assessments of sexual practices and orientations are especially prone to astonishing inconsistencies as well as cultural manipulations, unfounded biological foundations and emotional idiosyncrasies. But even though it may seem to many people that gay sex, for instance, is just obviously morally wrong, there are serious challenges to this judgment and many others. To the extent that these have not been met, the judgment has to be given up. This is, in fact, the opposite of a conservative picture. The *challenge and response* model can thus meet the conservativism objection on its own terms.

9 Moral Justification from an Empirical Perspective

I have argued that one of the main claims made by rationalism about the psychology of moral judgment is that there is a causally effective connection between moral reasoning and moral judgment. More precisely, the idea was that the reasons a subject can cite for her moral judgment also figure—in some way—in true causal explanations for why she holds that judgment. Haidt's SI model poses a challenge to the empirical accuracy of that claim—a challenge to whether it is *true* that there is such a connection—on the grounds that most of our moral reasoning comes after the fact. And because cause must be temporally prior to effect, such *post hoc* reasoning cannot be the cause of one's moral judgments. In response to this problem, I have distinguished between two different kinds of *post hoc* reasoning, one good and one bad. In the bad case, subjects confabulate: they make up entirely inaccurate rationalizations for their hopelessly indefensible emotionally triggered intuitions. In the good case, subjects do not consciously reason toward their moral verdicts, either; but they do make explicit the reasons that really did figure in the acquisition, formation, and maintenance of their moral intuitions. This type of reasoning is *post hoc*, but not confabulatory. And the simple fact that subjects do reason in such a way shows that they feel some sort of metapressure to be rational. They consider themselves to be under the obligation to hold only those moral judgments which could be justified with appropriate reasons.

The *challenge and response* model of moral reasoning I have developed and defended in this chapter is supposed to cash out the idea that there can be varieties of *post hoc* reasoning which do have an effective connection to subjects' moral judgments, thus reconciling the empirical evidence for the *post hoc* thesis with our most basic normative intuitions regarding the force of moral reasoning. After all, the most central idea behind the *challenge and response* model is that moral intuitions enjoy presumptive justification that does not have to be earned through conscious before the fact reasoning, but through meeting relevant objections. The episodes of explicit reasoning which consist in responding to the relevant challenges there are to one's moral intuitions are paradigmatically *post hoc* and even, to a certain extent, biased: they are supposed to *defend* the intuitions one already has, albeit in a reason responsive way. Moreover, these episodes will have a distinctively social character, because challenges are always encountered in some (possible or actual) social context, and reasoning always consists in responding to the audience who brought up the challenge.

But, as defendants of the SI model will insist, the dumbfounding experiments show that regardless of how we conceptualize genuine moral reasoning, that is, regardless of whether we think about it in terms of before the fact reasoning or reasoning in response to legitimate challenges: on both notions, reasoning turns out to be ineffective. What Haidt and his colleagues have done in their experiments is to test for both types of reasoning, and what they found was that people neither employ conscious reasoning to arrive at their judgments, nor respond adequately to the challenges which are put forward to their judgments by an interlocutor.

In all likelihood, this is to some extent correct. But that subjects do not respond *adequately* to the experimenters' attempts to debunk their gut feelings about incest, cannibalism, or other offensive yet allegedly harmless scenarios, remains an ambiguous observation. As I have already remarked, the social intuitionist experiments fail to distinguish properly between two kinds of inadequacy (Jacobson 2012): one that consists in mere *inarticulateness* when it comes to citing reasons for one's moral verdicts, and one that, in cases where there are objectively no reasons for someone's judgment, can rightfully be described as genuine dumbfounding. The former case is compatible with subjects having a *sensitivity* for the good moral reasons for or against a proposed course of action, even though they might not be able to make explicit what those reasons are. In fact, it is part of the job of the experimental devil's advocate to insinuate that only a rather narrow set of considerations—physical harm that results directly from the described action—count as valid reasons, thereby deliberately undermining ordinary subjects' confidence in their everyday moral competence. Of course, if one takes a closer look at the scenarios that have been used to make the social intuitionist case, one immediately sees that there are plenty of good ethical reasons not to engage in risky sexual acts just for the fun of it.

Now what about the claim that subjects do not respond adequately to legitimate challenges? The immunity of people's automatic moral intuitions to reasoned change has been grossly overstated. In Haidt's and Murphy's original experiment, the devil's advocate tries to persuade the respective participants that the siblings who have sex with each other are not doing anything wrong, by pointing out—in terms of objective act-consequentialist criteria—that there was actually no harm done. Most subjects' intuition stays the same, however, which has led to the social intuitionist interpretation that reasoning is ineffective in the formation of moral judgment. The problem with this conclusion is that participants' tenacity in holding on to their judgments might have more to do with the fact that they simply remain unconvinced by the experimenter's act-consequentialist reasoning, but do not have the resources or training to point out its flaws.

Another problem with this is that the moral rationalist does not want to say that such *bad* moral reasoning should have an effect on moral judgment. But do people change their minds when *good* moral reasons are presented to them? Are they happy to revise their intuition in the light of *legitimate* challenges? In order to answer this question, Paxton, Ungar, and Greene (2012) designed an experiment to test how people's moral judgments about the *Incest* vignette would be affected if they were given a good argument against the wrongness of consensual incest and enough time to reflect on it. When subjects change their mind in response to a bad argument, this could hardly be called proper reasoning. And when subjects change their mind in response to a good argument whose quality they did not have time to appreciate, this change could be attributed to nonrational forces as well. Thus, participants were randomly assigned to four different conditions (strong/weak argument × delayed/immediate response) in which they had to judge—on a scale from one to seven—the moral acceptability of Julie and Mark's action. Subjects were expected to intuitively disapprove of the described action they had to consider. In one condition, they were given an additional two minutes to think about a legitimate undercutting challenge to their intuitive judgment, which said that feelings of disgust toward incest probably evolved to prevent the birth of handicapped children, but given that Julie and Mark used contraception, this rationale does not apply anymore in the case at issue (Paxton, Ungar, and Greene 2012, 170). In another condition, they were given a really bad rebutting challenge to their intuition, which said that if siblings were not supposed to make love, then first, they would not be sexually compatible at all, and second, that having sex is an act of love—and the more love in the world, the better. What they found was that subjects' moral acceptability ratings went up significantly when, and only when, they were given the strong argument and enough time to reflect on it. This comes as close to an *experimentum crucis* for the SI model as possible, and it suggests that our moral intuitions are amenable to genuine reasoning.

Obviously, this susceptibility to moral reasoning might often be severely diminished, especially when people are in the grip of a very strong emotional reaction. But, the extent to which moral judges are simply slaves to their passions is exaggerated, and doesn't hold in the face of evidence. In a recent study, Matthew Feinberg and his colleagues (2012) could show that subjects were able to overcome their intuitive gut reactions by reappraising their emotional response. They found that people who habitually reevaluate their emotions as well as subjects who, though they might not have a steady disposition to do this, happened to reappraise their immediate reaction toward Haidt's *Incest* scenario and others, judged the described actions significantly less harshly (this was determined on the basis of blindly coded self-reports). In a third study, the researchers directly manipulated subjects' tendency toward emotional reappraisal, which had the same effect.

To be sure, there are cases of genuine moral dumbfounding and confabulation. One should not overstate the claims of rationalism and say that all subjects are always justified in holding all of their moral beliefs. Rationalism must not lose its critical bite. There is plenty of room for error, bias, and emotional distortion in the *challenge and response* model. I have started from the idea that a comprehensive theory of moral judgment must not only say something about how people judge, but also about how people judge correctly. But where there is correctness, there must also be (the possibility of) *in*correctness. Yet so far, I haven't said much about what it is to make mistakes when one judges morally, or what it means to reason improperly. Very coarsely put, the main mistakes people make when reasoning about moral issues are the

(1) failure to consider any challenges at all (closed mindedness),
(2) failure to appreciate the force of a particular challenge,
(3) failure to respond adequately to a challenge, and the
(4) failure to manage/monitor one's intuitions/judgments in response to the challenges one has received and the responses one has given.

This typology gives a better interpretation of the phenomenon of moral dumbfounding. As Haidt sees it, the phenomenon poses a radical challenge to the idea of moral reasoning. It suggests that it is always ineffective. With the above account in hand, however, we can see the following: Haidt is correct in saying that something goes seriously wrong in his dumbfounding cases. But what goes wrong there does not generalize to the very possibility of genuine moral reasoning.

How can this be applied to empirical psychology of moral judgment and reasoning? The main empirical theories of moral judgment and moral reasoning all have something to say about the relation between emotion, intuition, moral beliefs, and conscious reasoning. The DP model, for instance, holds that the distinction between automatic and controlled cognition can be mapped onto different types of moral judgment: deontological judgments are based on emotionally charged, primitive intuitions, while consequentialist judgments are based on controlled reasoning and deliberation. Therefore, only consequentialist judgments satisfy the requirements which are necessary for a judgment to be justified. Deontological judgments do not satisfy those requirements, and are thus not justified. Figure 3.1 shows a schematic representation of the model.

The SI model generalizes the claim the DP model makes about deontological judgments to all moral judgments. While the DP model holds that deontological judgments are based on emotional reactions, and that elaborate normative theories such as deontological ethics are confabulatory

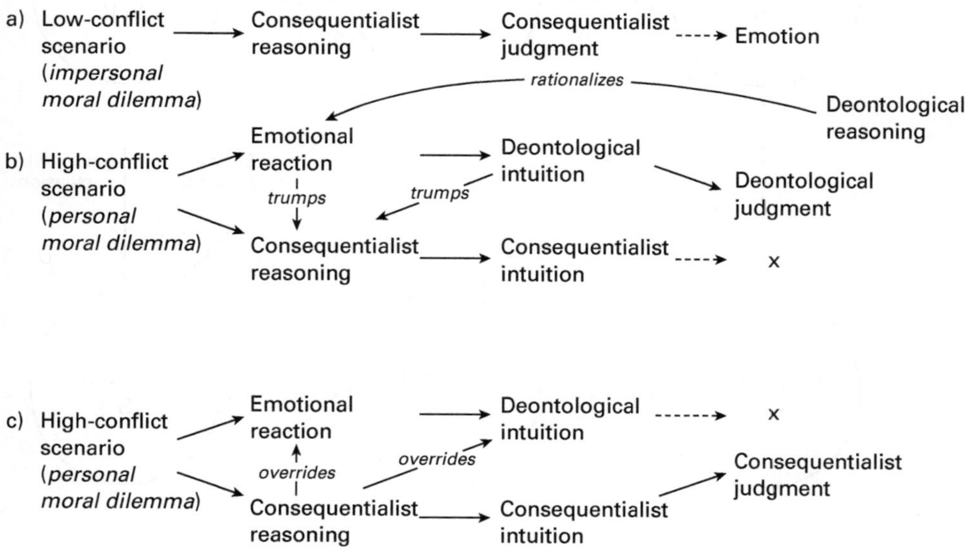

Figure 3.1
The dual process model of moral cognition.

after-the-fact rationalizations of those emotional reactions (see figure 3.1b), the SI model, says that this actually holds for all moral judgments. The DP model says that there are two tracks that lead to moral judgment: the emotional track leads to deontological judgments, and conscious reasoning in support of those is confabulatory. The cognitive track runs in the opposite direction: controlled cognition leads to consequentialist judgments, and emotional reactions follow in its wake. Subtleties aside, the SI model holds there is only the first track, which is responsible for all moral cognition. (The second track is supposed to be very rare at best.) Figure 3.2 shows Haidt's graphic representation of his theory (Haidt 2001, 815).

Let me now decompose this model into its elementary parts, and then put it together again from the perspective of the *educated intuitions* account of moral judgment and the *challenge and response* account of moral

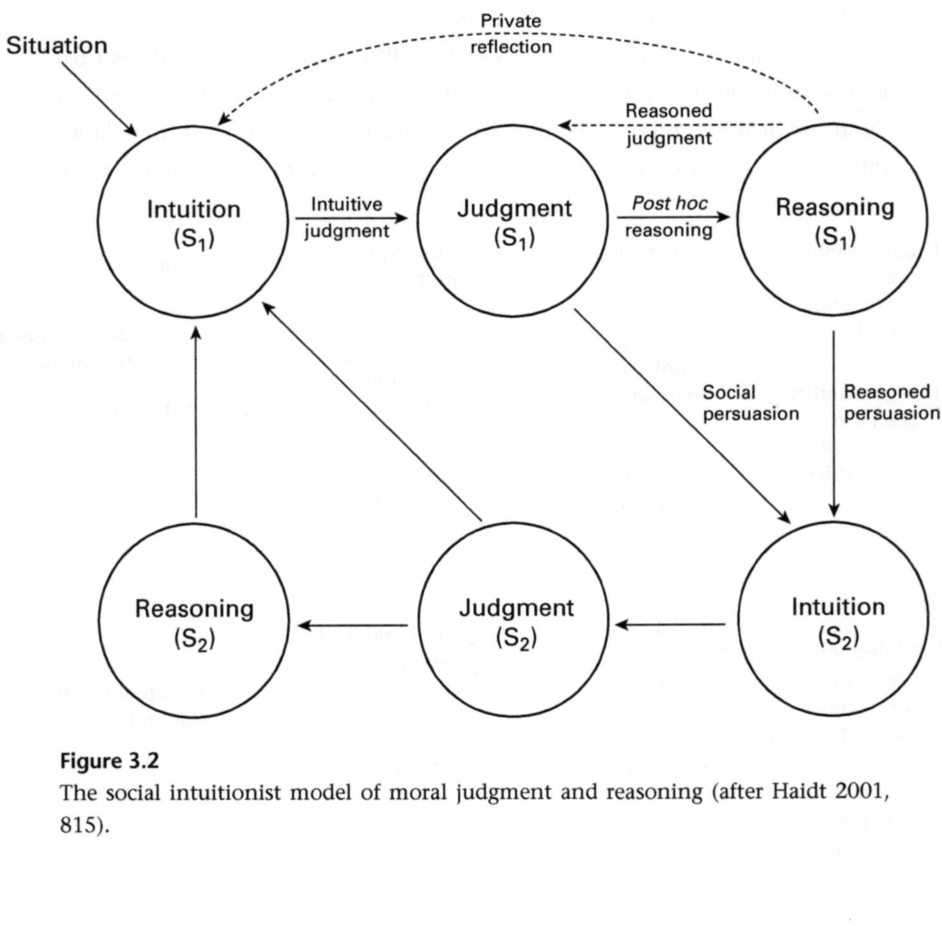

Figure 3.2
The social intuitionist model of moral judgment and reasoning (after Haidt 2001, 815).

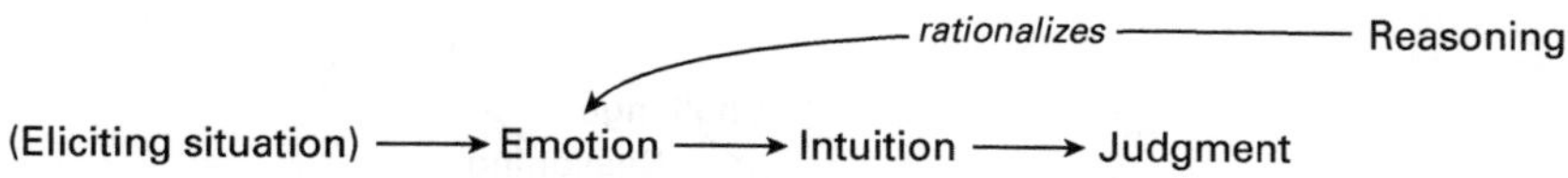

Figure 3.3
The *post hoc* link.

Figure 3.4
Simple rationalism.

reasoning. I take this to be the core of the model (figure 3.3). In Haidt's SI model, this corresponds to link (2). Emotional reactions trigger moral intuitions on which moral judgments are based; moral reasoning comes after the fact, rationalizing the irrational emotional reaction.

What the DP and the SI models agree on is that for a moral subject to be justified in holding a certain moral belief, the subject must have actively *earned* this entitlement with prior *before the fact* reasoning. That is why the processes depicted in figure 3.1b and figure 3.2 are said not to convey justification on their respective judgmental outputs. The DP and SI models *disagree* merely with respect to the question of whether subjects sometimes do earn these entitlements—for example, when they are making utilitarian moral judgments (as the DP model has it—figure 3.1a and 3.1c)—or whether they never—or only very seldom—do (as the SI model says).

But as we have seen above, we are given no convincing reason to accept this picture of moral reasoning to begin with. Figure 3.4 shows what the DP and the SI model assume the rationalist account of moral judgment is committed to, and what—according to this account—a process of genuinely effective moral reasoning would have to look like.

But we have no reason to think that his how subjects arrive at their judgments even in nonmoral cases. What does the *challenge and response* model say about all this? How does the account of moral reasoning just presented here tie in with the empirical story told in previous chapters? The first point to emphasize here is that moral reasoning is a challenge-dependent responsive practice, which leads to a different interpretation of the *post hoc* thesis and a rejection of simple rationalist models as depicted in figure 3.3 (see figure 3.5).

Figure 3.5
Challenge and response reasoning.

The next step is to realize that many current theories of moral cognition suffer from what can be called *focal bias*: by focusing on what goes on in *one* person's mind at *one* point in time, these models suffer from a severe form of temporal and social myopia (Gerrans and Kennett 2010), a blindness for the past and future of competent moral agents as well as the supraindividual character of the practice of moral reasoning. Most empirical models of moral cognition discount a person's past (the source of her intuitions and how she acquired their content) and future (what it means for a subject to engage in the critical practices that define sapient beings). The *challenge and response* model avoids these mistakes, by offering an account of the education of our intuitions and a description of the social structure of moral reflection.

The second point is that typically, conscious reasoning toward a moral judgment is not required for that judgment to be based on moral reasoning, because moral intuitions are malleable: they are not eternally fixed, brutish, and cognitively impenetrable responses, but subject to continuous improvement, overlearning, and habituation. This is what I have referred to as the education of our moral intuitions. I have already described what types of education there are, and why we have good reason to think that nothing about the rationality of a cognitive process needs to change simply in virtue of the fact that it has become automatic over time. I have also argued that explicit moral reasoning and deliberate reflection is one means by which our intuitions are educated. In this chapter, I have spelled out in more detail what I take the structure of moral reasoning to be, and what it means to say that moral reasoning is best seen as a practice of responding, in various ways, to legitimate challenges of various types. If we "zoom out" a little further, thereby integrating the perspective of the education of moral intuitions, we can redescribe *post hoc* reasoning as a reason transmitting feedback cycle (figure 3.6).

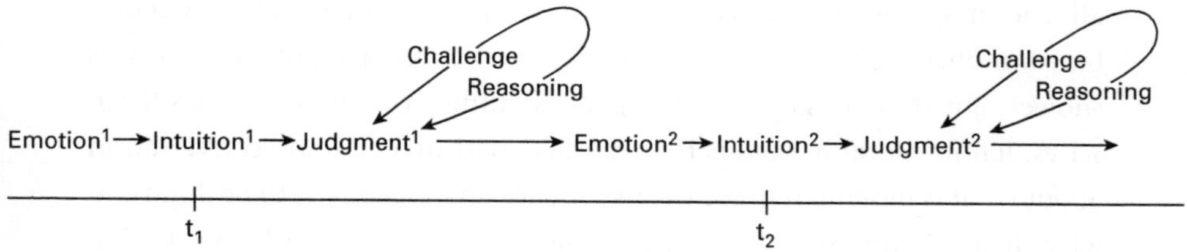

Figure 3.6
The educated intuitions model.

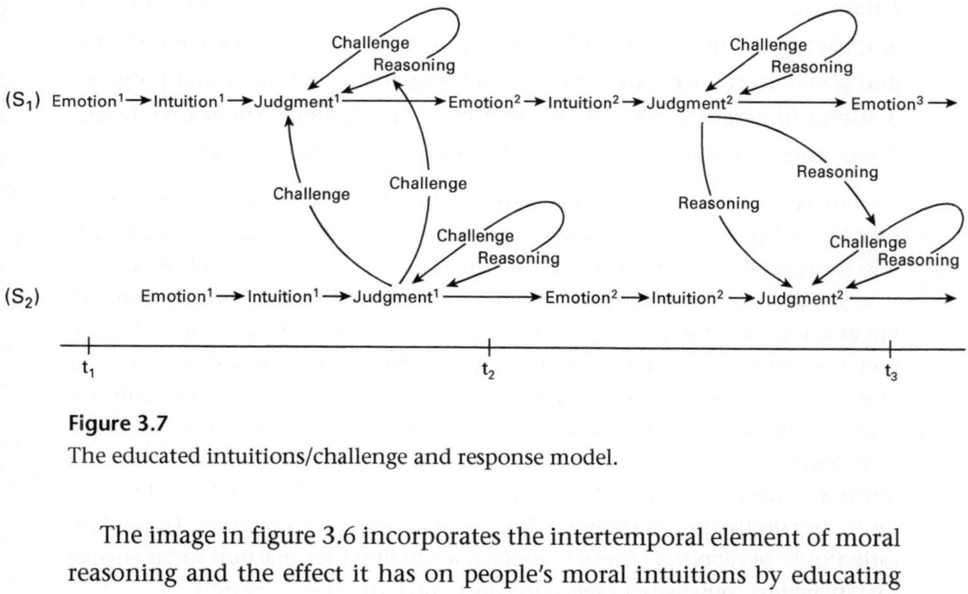

Figure 3.7
The educated intuitions/challenge and response model.

The image in figure 3.6 incorporates the intertemporal element of moral reasoning and the effect it has on people's moral intuitions by educating them. But without the social element, the model is still incomplete, which is why an interlocutor has to be introduced into the representation; we zoom out one more time and get what is shown in figure 3.7. This model is supposed to capture the essentially *diachronic* and *social* dimension of moral reasoning. Moral reasoning is a constant temporally and intersubjectively extended process of intuitive judgment formation and motivated reasoning, which leads to a permanent, intersubjectively entertained improvement of one's intuitions, improvement of one's reasoning and improvement of the interaction of intuitive and controlled cognition.

It should be noted that Haidt's SI model only *seems* to avoid the mistake of focusing on only one morally judging subject. Isn't it called *social* intuitionism, one might ask? But in fact, social intuitionism isn't very social at all: the moral agent as described by the SI model is a radically unencumbered, monadic subject. Of course there are other people out there for this subject. But they never come into play as entities which are *subjects themselves,* fellow moral reasoners that can take part in a genuine conversation about what is morally right or wrong, and about what ought to be done and why. Instead, the SI model misdescribes them as potential objects and sources of strategic persuasion, never as persons who can put forward reasons which are worthy to be considered.

Another virtue of the educated intuitions/challenge and response model I have developed here is that it provides a normatively convincing and empirically adequate schematic description of the way moral judgments and social moral reasoning interact to bring about real-life moral progress. A useful illustration for this can be found in Elizabeth Anderson's recent discussion of moral progress toward the abolition of slavery. She writes:

> By contrast, moral reflection is merely speculative when there is no particular practical need to deliberate brought about by vivid confrontation with the actual conflicting claims of others, or the bad consequences of current norms, or doubts about how to apply given norms. People may entertain hypothetical challenges to a norm, but not as voiced by actual people complaining about the status quo and calling its supporters to account for their conduct. Few people find purely speculative individual moral reflection motivating. Because moral norms purport to have authority for everyone, and aim to coordinate everyone's normative expectations, how can one be confident that one's own moral reflections are on the right track unless one sees that others are prepared to orient their conduct in accordance with them? Moral reasoning, to be effective in changing social practices, must be done together. The mobilization of large numbers of people supplies a condition for practical social change that individual conscientious reflection does not. (2014, 14)

In line with the general pragmatist flavor of the model developed here, Anderson contends that people do not achieve motivationally effective moral insights by gradually becoming better at applying certain foundational moral principles to the world. Rather, moral progress works through improvement from within. Most importantly, this involves identifying and counteracting systematic biases that may operate on one's moral judgment, as well as incorporating proposed revisions of one's moral outlook into experiments of living to test their practicability outside the armchair.

One further advantage of my account I would like to mention is that, unlike other rationalist models of moral judgment that place steeper demands on subjects' reflective capacities, the educated intuitions/

challenge and response model affords center stage to the notion of "reflective triage" (Doris 2015, 73–77). People cannot reflect about what to do all the time; reasoning has to be "judiciously applied at appropriate junctures" (Doris 2015, 73). The model developed here offers an account of where those junctures are, and thus how to square people's limited resources of reflective intervention with the fact that such intervention is frequently required.

The educated intuitions/challenge and response model is a form of "default interventionism" (Evans and Stanovich 2013, 236). The general structure of cognition is such that intuitive modes of processing generate a default response whose validity is monitored by reflective types of processing and, if necessary, overridden. However, it somewhat departs from more traditional accounts of intuitive override in that it does not primarily delegate the task of override initiation to internal mechanisms of conflict detection (such as a "feeling of rightness," see Thompson, Prowse Turner, and Pennycook 2011), but to externalized forms of socially conducted reasoning.

Some will certainly remain unconvinced by the praise I heap upon moral reasoning as a social practice, and insist that the costs of rationalization and other forms of biased reasoning outweigh its benefits even in the kind of competitive marketplace of ideas I describe above (Schwitzgebel and Ellis forthcoming). To address this worry, let me remind you again what those benefits are:

> Reasoning is best employed in a dialogical context. In such contexts, opening a discussion with a relatively weak argument is often sensible: It saves the trouble of computing the best way to convince a specific audience, and if the argument proves unconvincing, its flaws can be addressed in the back and forth of argumentation. Indeed, the interlocutor typically provides counter-arguments that help the speaker refine her arguments in appropriate ways. ... As a result, the laziness of argument production might not be a flaw but an adaptive feature of reasoning. By contrast, people should properly evaluate other people's arguments, so as not to accept misleading information. (Trouche et al. 2015)

So much for the *efficiency* gains associated with socially extended reasoning. However, the *epistemic* case for moral reasoning as a social practice is equally strong. In an intriguing recent study that combines an argumentative account of moral reasoning with a choice blindness design, Trouche and his colleagues found that the *quality* of the assessment of people's own arguments in favor of a particular conclusion improves significantly when subjects are led to believe that they are evaluating someone else's reasoning. The social practice of moral reasoning, even when entertained by a community of "defensive" moral reasoners, constitutes an attractive trade-off between deliberative cost-effectiveness and epistemic rigor.

In a certain sense, then, it is actually incorrect to say that moral judgments are based on educated intuitions, just as it is incorrect to suggest that they are based on emotional reactions, beliefs about moral facts, or what have you. That is because narrowing the perspective in such a way suggests that there could be such a thing as a single moral judgment, based on a single moral intuition. But that is not the case: individual moral judgments depend, for them to be possible at all, on an entire practice of moral cognition that extends over individual points in time as well as individual moral agents. The study of moral judgment has to start from the idea that moral thinking is deeply entrenched in a continuous stream of intertemporal and intersubjective moral reasoning. Moral judgments, then, are the individual verdicts arrived at by those competent moral subjects who participate in this social practice. They are based on emotionally charged moral intuitions which are subject to diachronic education and the reflective criticism put forward by one's fellow moral reasoners. The question whether moral judgments are based on emotion, or reason, or both, cannot be addressed only by looking at the *mental faculties* involved in producing them. The study of moral judgment is always the study of that whole peculiar institution "morality."

Conclusion

This chapter concludes my constructive response to the anti-rationalist challenge. The challenge had two parts: one was about the automatic, intuitive character of moral judgment, and the fact that moral reasoning is a *post hoc* enterprise. The other was about the social dimension of moral reasoning.

I have argued, first, that the claim that moral reasoning bears an effective connection to people's moral judgments can be reconciled with the automaticity of those judgments because over the course of our moral education, patterns of reasoning migrate into our intuitive responses. Moral reasoning figures in the acquisition, formation, maintenance, and the reflective criticism of our intuitions.

Second, I have argued that this account of moral judgment is nicely complemented by a model that describes moral reasoning as a social practice. This challenge and response model argues that in their moral reasoning, subjects always start from their moral intuitions. This explains, from yet another angle, why moral reasoning comes after the fact. Moreover, subjects only start to reason when they encounter a legitimate challenge to their beliefs—a reason, put forward to them in a real or imagined social discourse. I have used Mark Timmons's structural moral contextualism as an example of how to understand the fact that it is often epistemically responsible to accept one's moral intuitions as basic, that is, as not standing in need of justification. This means that Haidt charge against the nonrationality of not consciously justified moral intuitions misses the point. Moral intuitions are susceptible to conscious reasoning not because they are preceded by it, but because they are open to reasoning in the light of legitimate challenges. I have explained what kinds of challenges and responses there are, and that the SI model is correct in saying that people often fall short of their epistemic obligations. The claim, however, that regardless of how moral reasoning is conceptualized, it turns out empirically that subjects' moral intuitions remain impenetrable, can be rejected. I have provided empirical evidence for the fact that under moderately improved epistemic conditions—that is, sufficient time to reflect on a valid challenge—moral intuitions are responsive to explicit reasoning. Finally, I have used the main tenets of this account to offer a reinterpretation of moral reasoning as a social practice, in which morally competent agents challenge each other's educated intuitions and respond to each other's challenges. This intersubjective chain of improved moral judgments and justification, I wish to maintain, is the source of individual and social moral progress.

II The Emotionist Challenge

Do you want to meet tomorrow or outside? This question sounds odd, because it rests on a confusion: here are two distinct conceptual spheres that do not penetrate each other, and the question presents them as a misleading dichotomy. It makes the erroneous suggestion that there is an exhaustive alternative that one has to choose from; and it falsely insinuates that by choosing one of the alternatives, one has rejected the other. In what follows, I shall assume that the dichotomy between rationalism and emotionism (or, more traditionally, sentimentalism) about moral judgment rests on a similar confusion.

Rationalism is about whether there are any moral judgments that are, in some sense yet to be specified, "better" than others—more justified, more correct, more rational. Moreover, it is about whether moral progress is possible, and whether the practice of moral reasoning is a sham, or something that ought to be taken seriously. Emotionism, on the other hand, is a theory about whether and to what extent emotions are involved in moral judgment. It is about the psychological foundations of moral thinking, and about what mental capacities human beings recruit in order to arrive at their moral verdicts. As such, it has nothing whatsoever to say about whether there are any normative standards with which we can legitimately assess the output of those capacities. We can be rationalists *and* emotionists about moral judgment, just like we can meet tomorrow *and* outside.

The idea that emotions are not brutish irrational forces that lack intelligence has now become more or less uncontroversial among psychologists and empirically minded philosophers (Jones 2006). It is all the more puzzling that people still announce the demise of moral rationalism on the basis of evidence suggesting that emotions are essential for moral judgment. Jonathan Haidt, one of the foremost champions of the anti-rationalist strand in moral psychology, argues that rationalism about moral judgment is false because (most) moral judgments are based on emotionally charged intuitions. But at the same time, he also writes: "Emotions are not dumb. ... Emotions are a kind of information processing" (2012, 45). This entails that moral judgments, according to his claim, are based on something that processes information in a nonstupid way. What more could a rationalist ever ask for? Sharon Krause elaborates on this theme:

> In fact, sentiments are part of practical rationality itself. There is no faculty of practical reason that entirely stands apart from sentiment. Among other things, sentiments set the basis for future decisions by providing a sense of what matters, based on prior learning and experience. In other words, sentiments constitute the horizon of concern within which practical judgment and deliberation transpire. (Krause 2008, 3)

If this is true, and if the thing that rationalists about moral judgment ought to care most about is the *normative quality* of our moral judgments rather than on the basis of which capacities they are arrived at, then the emotionist challenge does not give rationalists much to worry about. If emotions are part of our practical rationality, why would the emotional nature of moral cognition undermine moral rationalism?

Many rationalists nonetheless feel attacked by emotionism. In doing so, they are committing the same conceptual mistake emotionists and sentimentalists are guilty of when they present their *psychological* claims as attacks on rationalism's central *normative* tenets. I can only speculate about this, but my suspicion is that there are (at least) three ideas that keep this apparent conflict between rationalists and sentimentalists alive:

(1) *Faculty psychology*. Modern philosophical thinking since Descartes operates within a mentalistic vocabulary that identifies reason and emotion with two distinct mental faculties. Only reason gives us ideas that are *clare et distincte*; emotion and feeling are, as Leibniz and others have put it, a form of *cognitio confusa*. Whether we like it or not, we have inherited some of the implicit assumptions of this vocabulary and thus tend to think of rationalism and sentimentalism as mutually exclusive alternatives. But if one drops the faculty psychology, which we have every reason to do, this stark opposition makes little sense.

(2) *The threat to autonomy*. The emotional basis of morality seems to pose a threat to us as autonomous moral agents. We cannot choose our feelings: they force themselves upon us, they overwhelm us, they are beyond our control. Autonomy and the passive nature of emotions thus seem to exclude each other. But if that is the case, and if moral judgments are based on emotions, and if these judgments, in turn, determine the morally significant choices we make as agents, then what choices we end up making is determined by things that force themselves upon us, overwhelm us, and are beyond our control. Therefore, our moral judgments are not made autonomously.

This idea surely has more to commend it than an outdated form of faculty psychology, but I doubt that it gives us much to worry about. First of all, unless one wants to subscribe to an implausible form of direct doxastic voluntarism—according to which subjects are autonomous with respect to their judgments because they can directly choose which propositions to believe (Williams 1970)—the same threat looms for nonmoral beliefs as well. Indirect doxastic voluntarism is less controversial, but people typically do have indirect control over their emotions, so there is no specific threat to rationalism, either. Second, even if subjects had no form of voluntary control over their beliefs, this would only entail that they lack judgmental autonomy if voluntarism were a necessary condition for autonomy. But it is far from obvious whether being able to voluntarily control which judgments one holds would be desirable in the first place: when it comes to making moral or nonmoral judgments, we are interested in acquiring only those beliefs that are true, rather than those we please to have. Being passive is not a problem as long as one's judgments are reason responsive.

(3) *The threat to morality*. Finally, the emotional basis of moral judgment seems to pose a threat to morality itself. Indeed, this third threat seems to directly follow from the purportedly heteronomous nature of moral cognition. If moral judgments are grounded in emotion, this seems to render the universality of moral norms impossible. Moral norms ought to apply to everyone equally, and for this to be possible people must be equally capable of gaining insight into what morality requires—that is, of making moral judgments. But if moral judgments are based on utterly contingent emotions, then this universality cannot be guaranteed, because

emotions are fickle and culturally relative. However, it is far from obvious whether emotions really are that contingent: most dispositions to experience morally relevant emotions such as empathy or guilt are widely shared. And even if emotions were contingent in this sense, so are nonmoral beliefs. Which descriptive beliefs we end up having depends on the perceptions the external worlds bestows upon us or the environment we grow up in.

These are not fully developed arguments, and they are not supposed to be. They are merely supposed to make the (false) idea that emotion and reason are arch enemies somewhat *understandable*. I have suggested that this claim should be rejected. But I do not wish to leave it at this mere reassurance. Instead, I now wish to provide some preliminary evidence for the complexity of the relationship between emotion and (moral) judgments. When it comes to morality, emotions are neither intrinsically unreliable nor infallible. And the absence of emotion neither guarantees that the quality of our moral judgments goes up, nor that it does down:

(1) *Emotions sometimes improve our judgments*. In the Wason selection task, test subjects are given a set of four cards and have to test a rule by selecting which cards to turn over. The rule employs a conditional such as "If there is a vowel on one side of the card, then there is an even number on the other side." On the visible side of the cards are a vowel, a consonant, an even number and an odd number. Most people turn over the cards with the vowel and the even number to test the rule. But this is a mistake, of course, as the rule does not say that there can be no consonant on the other side of an even number. To really test the rule one has to turn over the vowel card and the odd-number card.

However, when this abstract rule is replaced with a rule that specifies a social norm, test subjects perform much better. Leda Cosmides (1989) did a variation of the selection task in which subjects had to test a rule saying "If a person is drinking beer, then he must be over 20 years old"; participants received cards that said drinking beer, drinking coke, 25 years old, and 16 years old, respectively. In this case, most people turn over the right cards: the beer drinking and the 16-years-old card. People's quick, emotionally charged intuitive reaction toward the breaking of a social norm significantly improves their reasoning in this case. Examples from the philosophical literature about how a subject's emotional reactions can improve her moral decision making include, among many others, Nomy Arpaly's (2000, 2003) cases of inverse akrasia. As far as his consciously endorsed moral principles go, Huckleberry Finn thinks it right to turn Jim in. But his feelings of friendship toward the runaway slave make him unable to do so.

(2) *Emotions sometimes interfere with our judgments*. It would be naive to suppose that emotions are always the most reliable guides. Consider the famous "Asian disease" case (Tversky and Kahneman 1981). In this experiment, two groups of participants are given a choice between two government programs to fight a dangerous disease. In both conditions, it is estimated that the disease will result in six hundred deaths. Participants have to choose between program A and program B. In the first condition, A will save two hundred of those six hundred people; if B is implemented, there is a 1/3 probability that six hundred people will be saved, and a 2/3 probability that everyone will die. Here, most people prefer option A. In the second condition, there is program C, which will lead to the death of four hundred people, whereas if program D is adopted, there is a 1/3 probability that no one will die and a 2/3 probability that everyone will die. A and C are clearly the same, as are B and D. Yet most participants prefer A over B and D over C. The way these options are framed trigger differential emotional responses in people's minds, and these responses cloud their judgment.

(3) *Absence of emotions sometimes improves our judgments*. Consequentialists think that only consequences determine whether an action is right or wrong. For the sake of the argument, let's assume that you agree. And let's also assume that this means that you approve of pushing a fat man off a bridge in order to save five people from a runaway trolley.[1] It turns out that most "normal" people do not agree with you (Greene et al. 2001), but psychopaths typically do (Koenigs et al. 2007; Bartels and Pizarro 2011; see also Cima, Tonnaer, and Hauser 2010).[2] Psychopaths are known to suffer from emotional impairments. Usually, this also impairs their capacity to make moral judgments, and to draw certain important moral distinctions (Blair 1995). But in this case, lack of emotional resonance makes them more reliable, and improves their performance in the *Trolley* task. Now this example will of course seem objectionable to some, namely nonconsequentialists, or consequentialists who disapprove of pushing the man. But it shows that whether a lack of emotion improves or undermines one's ability to make moral judgments is a genuinely open question that has to be decided on a case by case basis.

(4) *Absence of emotions sometimes interferes with our judgments.* On the other hand, consider Antonio Damasio's (Saver and Damasio 1991; Damasio 1994) studies with patients who have suffered damage to the ventromedial prefrontal part of the brain. These patients also suffered from impaired emotion, and seem to have difficulties generating the "somatic markers" (as Damasio calls them), that quickly and subconsciously indicate whether a given behavioral option is desirable or not. People's performance on the *Iowa Gambling* task, for instance, shows that when people suffer from diminished or impaired emotional responsiveness, they lose their ability to make prudent, practically rational decisions. When having to choose a deck of cards to pick from, these people lose their ability to identify the better decks. Healthy people, it seems, have that ability, and are able to exercise it before and without having any conscious knowledge of what the best decks are. These are the examples that Sharon Krause (2008) has in mind when she emphasizes that there seems to be no faculty of practical reasoning that can function properly without the assistance of sentiments. Examples from the philosophical literature concerning how an absence of appropriate emotional reactions can undermine one's capacity to make good moral judgments are legion (Singer 2005).

All of this taken together shows that the relationship between emotion and sound judgment withstands a simple characterization. Moreover, the claim that emotion—or lack thereof—sometimes interferes with or improves people's (moral) judgments and decisions does not yet specify in any detailed way how this interference and/or improvement is supposed to go about, that is, of what type the relationship between feeling and morality is and how strong it has to be such that the above effects can be accounted for. Here are some ways in which emotion and moral judgment are connected, regardless of whether this connection is advantageous or detrimental:

(1) *Emotions are correlated with moral judgments and decisions.* When people make moral judgments, they are in an emotionally engaged state of mind. Rationalist or not, it would be unnecessary for anyone to deny this truism. This also means that the claim is too weak to have any bearing on the nature of moral judgment. A nutty professor can be so passionate about her formulas that whenever she calls them to mind, she feels a pleasant tingling sensation in her stomach. This does obviously not entail any form of sentimentalism about mathematics.

(2) *Emotions causally influence moral judgments and decisions.* We all know from everyday experience that our judgments covary with our feelings. My neighbor parked his car in front of my driveway, and I cannot get out. When in a calm state of mind, I might judge my neighbor's behavior less harshly. But when I am angry or annoyed about something, or just in a bad mood, I might easily feel tremendous outrage upon the impertinence of the cretin next door.

(3) *When making moral judgments, people often introspect their emotions.* Imagine a friend telling you that she attended a mutual acquaintance's wedding gown fitting. Your friend thought the dress was dreadful: unflattering, tasteless, and tacky. But she lied to the bride-to-be, telling her that she thought it was wonderful: flattering, tasteful, and classy. How do you figure out whether her behavior was OK? Do you browse your repertoire of moral principles, find out which pertain to the matter at hand, and then pass a verdict? Typically, this is not how things go about. What you will do in the overwhelming majority of cases is to listen carefully to your friend's story, take into consideration what you know about your acquaintance's character—particularly about how sensitive she is or how easily she can cope with criticism—and then settle on what "feels right."

What complicates things even more is that emotions themselves can be the objects of rational and moral assessment. We have seen that emotions can stand in a conducive or hindering relation to moral judgments by being correlated with or causally connected to them or by being consulted by moral judges when it comes to forming a judgment. But judgments can also be directly *about* emotions:

(1) *Emotions can be rationally assessed in terms of fittingness.* Certain emotions only apply to certain types of cases. Disgust applies to sticky, oozy, and contaminating things; sadness applies when someone has irretrievably lost something; guilt is only appropriate in response to one's own wrongdoing. It would not make sense to cut across these emotional boundaries.

(2) *Emotions can be rationally assessed in terms of their factual basis.* With only few exceptions (such as *Weltschmerz*) virtually all emotions depend on some kind of factual basis, and can be rationally criticized on factual grounds. I may be jealous about my colleague, whom I suspect of philandering with my wife; but when it turns out that it is not my colleague who

is guilty of this violation of my trust, but my boss, I will readily redirect my jealousy toward this new target. And when I find out that no philandering happened at all, my jealousy ought to vanish altogether.

(3) *Emotions can be rationally assessed in terms of their proportionality*. It is not just the quality, but also the quantity of my emotional reactions which can be inappropriate. Genuine grief cannot last just one afternoon; a case of murder calls for stronger resentment than shoplifting.

(4) *Emotions can be rationally assessed in terms of consistency*. The moral principles that, perhaps implicitly, figure in my emotional responses toward cases ought to be consistent with each other. It is irrational to condemn one instance of lying while being inclined to excuse other such instances for no obvious reason.

(5) *Emotions can be rationally assessed in terms of the self/other distinction.* If one feels ashamed about one's actions and, upon reflection, judges that this is appropriate, one ought to think that other people who find themselves in the same or a sufficiently similar situation ought to be ashamed as well. The same holds for other feelings and emotions: what makes guilt or fear appropriate for one person also makes it appropriate, *ceteris paribus*, for another.

(6) *Emotions can be rationally assessed in terms of their moral value*. Sometimes, emotions can become subject to moral assessment. Oftentimes people consider racist, misogynistic, or homophobic feelings to be disgusting in and of themselves, regardless of whether the person who has them endorses or acts upon them.

For a start, I will take emotionism to be the claim that emotions are somehow *essential* to moral judgment. More precisely, moral judgments are purported to be based on emotions in such a way that reason becomes utterly superfluous. All of this sounds rather obscure at this point. And this is no accident, because it is part of the execution of the task I have set myself to clarify what is meant by "emotion" and "reason," why they are allegedly incompatible with each other, what, accordingly, *emotionism* and *rationalism* might and might not mean, and in what sense moral judgment might be "based" on one or the other.

The relationship between morality, emotion, and reason is like most other relationships: it's complicated. The second part of this book will be about this relationship, and why a tight link between feeling and moral judgment poses no threat to rationalism about the psychology of moral judgment. I will explain what the emotionist challenge consists in, how it complements the anti-rationalist challenge and how the moral rationalist can respond to it. I will start with the problem of moral error, which those who embrace the emotionist challenge to rationalism about moral judgment are especially ill-equipped to account for. But let's not get ahead of ourselves.

4 Moral Error

Introduction

In a 1971 interview broadcast on Granada TV in Manchester, Woody Allen, in his trademark self-deprecating manner, said about one of his early films: "It was a boring picture, as I recall." The interviewer responded, surprised: "I rather enjoyed it." To which Allen replied: "Yes, but you're mistaken." In the world of humor, this reply sounds odd—which is why it is funny. In the moral domain, an exchange like this would not sound weird at all. What is or is not funny is settled by what we find funny. There is no being mistaken about it. Not so for morality: moral discourse allows for a considerable amount of error. Emotionism about moral judgment, I will argue in this chapter, cannot explain this fact.

The second part of this book is about the emotionist challenge to rationalism about moral judgment. I will argue that, in contrast to the first part, this challenge need not be overcome, because it can be dissolved. Once it is properly understood, the emotional basis of moral cognition poses no threat to the rationalist project.

In what follows later, I will discuss the most important empirical evidence there is for the claims that emotions are both necessary and sufficient for moral judgment, and I will present arguments that aim to show that, provided the right reading of them is put on the table, the truth of these claims need not undermine rationalism about the psychology of moral judgment. But, one might ask, if moral judgments are not merely triggered or accompanied by emotions, if feeling is essential to moral cognition, why argue for a rationalist position at all? Why not simply settle for emotionism?

In this chapter, I show why, evidence for the tight link between emotion and morality notwithstanding, we should still opt against an emotionist account of their relation. Here is the main gist of the argument developed in this chapter: emotionism entails that moral error is impossible. Usually, this is thought to result from the fact that it holds that moral judgments cannot be false *or true*, because they do not aim at truth to begin with. I wish to maintain that empirically informed versions of emotionism entail that moral error is impossible because moral judgments are *necessarily true*.

My diagnosis is based on what I call "the wrong kind of mistake" problem (henceforth, WKM problem). The problem, to put it bluntly, is that no plausible theory of moral judgment should have the implication that moral error is impossible. Moral judges should not come out as *morally* infallible. Emotionists accept this constraint but cannot satisfy it.

Unfortunately, putting sentimentalists where I want them is not as simple as it sounds, because robust sentimentalists are also *relativists*. As such, they are often happy to accept claims about moral error which most of us would find implausible, or at least unorthodox. For example, they typically have no problem admitting that two moral judges who seem to disagree about a normative issue really—that is, beneath the assertoric surface structure of moral discourse—need not disagree at all; likewise, they are happy to accept the claim that two people who have conflicting moral views can both be right. The fact that there is no such thing as being mistaken about the moral facts does not threaten robust sentimentalists—it's the very conclusion they're after. Because of this willingness to embrace the counterintuitive, I will try to motivate the sentimentalist to think again about moral mistakes on grounds that do not require her prior agreement that moral error is possible. To achieve this, I will work with the metaphysically more modest assumption that moral judges are able to *change their minds*.

Once I have pointed out that not being able to properly explain moral error comes with the additional cost of failing to understand changes of mind about moral matters, I can begin to elaborate on my main theme: although robust sentimentalists about moral judgment should *desire* to be able to account for moral mistakes, this desire must remain frustrated, because their theory does not leave room for the *right kind* of moral error. Here is why: like all theories of moral judgment, emotionism consists of three building blocks. Every theory needs to say something about the *attitudes* that are involved in moral judgment, the *objects* that these judgments are about, and the *information* that is relevant to the relation between these attitudes and their objects. Thus there are three points at which an account of moral mistake can be integrated into the theory. Moral mistakes can be made about the *feelings* moral judges have, the *objects* they respond to with their feelings, and they *information* they rely on in responding the way they do. None of these options gives the sentimentalist what she wants: an account of genuine *moral* mistake. All versions of this theory fall prey to the WKM problem.

This chapter has five sections. In section 1, I will briefly introduce the preliminary account of emotionism I will work with. I will add further refinements to the position in subsequent chapters, but for my purposes here, this rendition will suffice. Section 2 shows why this account is confronted with the WKM problem. First, I argue that there is yet another problem for emotionism—which I shall refer to as the *infallibility problem*[3]—that gives the emotionist very good reason to develop an account of genuine moral error in the first place. The WKM problem, then, arises when the emotionist attempts to give such an account. In section 3, I argue that biting the bullet and accepting that genuine moral error is impossible is not an option for the sentimentalist, because there are other, equally serious problems that come with this strategy. Section 4 goes through the three notions of moral mistake the sentimentalist *can* allow for, and shows that none of them captures the right kind of mistake, the kind we are interested in when we challenge our own moral beliefs and those of others head-on. In short: moral error is not about getting your own feelings, the objects of your feelings, or the facts wrong. Moral mistakes consist in making moral judgments which are incorrect in a distinctively moral way. Section 5 addresses an important worry one might have with regard to the argument of this chapter. It seems that my criticism is begging the question against emotionist theories of moral judgment because it demands an account of a form of genuinely moral error that emotionists think simply does not exist.

1 What Is Emotionism?

Emotions matter to moral judgment. This trivial claim is part of common sense as well as scientific theory, and no one, not even the most die-hard rationalist, would be well advised to deny it. But emotionism makes stronger claims. More precisely, I take it to be the conjunction of a *psychological* and a *metaphysical* thesis:[4]

Emotionism

PT. Moral judgments are constituted by emotions.
MT. Moral properties are constituted by emotions.

In subsequent chapters I will discuss emotionism as an account of the psychology of moral judgment. Making a moral judgment, this part of the account has it, is to have an emotional reaction toward a morally salient case. This psychological component is captured by PT. But a full-blown version of emotionism does not stop here: it also wants to say something about the nature of moral properties; and, perhaps unsurprisingly, it holds that these properties are constituted by emotional reactions as well. This second part is required for emotionism to be a comprehensive metaethical doctrine, because otherwise, one could argue that moral values are not constituted by emotions, but that emotions simply are how subjects gain epistemic access to moral facts. This is not what the emotionist is after.

Let me emphasize what other sentimentalist accounts of moral judgment there are to which my argument does *not* apply. First, there is neosentimentalism. Very roughly, this cluster of theories holds that to make a moral judgment is to judge it appropriate, warranted, or justified to have a certain emotional response. In this chapter, I am *not* talking about neosentimentalism. Second, there are what might be called traditional forms of sentimentalism, by which I mean the theories championed by Hume, Smith, and others. It is unclear whether these theories would be a target of my criticism, and I shall remain agnostic about that question. Although they do agree with emotionism about the fact that morality is, in some sense, based on emotion, they do not seem to make the above distinction between the emotional basis of moral *judgments* and the emotional nature of moral *properties*. (The same point, by the way, seems to apply to Nichols's [2004] account of moral cognition.) Therefore, I am *not* talking about traditional sentimentalism (although if the same distinction *could* be found in traditional sentimentalism, my argument would apply to it as well). Third, there is the recent empirical psychological trend to debunk so-called rationalist models of moral judgment on the basis of evidence that suggests morality is all about emotionally charged intuitions rather than reasoning. It is unclear whether the challenge I shall put forward applies to these models as well. I suspect that, if pressed, most psychologists would agree to both psychological as well as metaphysical emotionism about morality. If this speculation is correct, then my argument does apply to their respective theories.

Which authors can unambiguously be said to hold the position my objection applies to? The following discussion of the possibility of error on an emotionist account might give one the impression that my criticism applies merely to Jesse Prinz's (2006, 2007) so-called constructive sentimentalism. But this impression would be misleading, and is largely based on the fact that so many sentimentalist theories of moral judgment fail to specify clearly what their position is committed to. On the other hand, Prinz is the only author—at least to my knowledge—who clearly sees the need for sentimentalist theories of moral judgment to separate the *psychology* of value judgments from the *metaphysics* of evaluative properties. Other theorists tend to conflate the two; to see this conflation at work, take a look at the following passage:

> The great underlying agreement between the neosentimentalists, which is obscured by their metaethical dispute, is that they are all committed to what we will call the *response dependency thesis* (RDT). The crucial idea, which we take to be the defining characteristic of neosentimentalism, is that an important set of evaluative *concepts* (*or terms or properties* [emphasis added]) is best understood as invoking a normative assessment of the *appropriateness* (or merit or rationality) of some associated emotional response. Hence,
>
> (RDT) To think that X has some evaluative property Φ is to think it appropriate to feel F in response to X. (D'Arms and Jacobson 2000a, 729)

One can easily see that D'Arms and Jacobson treat the claims that evaluative concepts as well as evaluative properties are response dependent as interchangeable, even though the two amount to entirely different claims. Evaluative concepts are components of *judgments* about evaluative states of affairs, evaluative properties are components of those states of affairs *themselves*. It remains unclear whether neosentimentalists, for example, endorse a sentimentalist account of both moral judgments *and* moral properties. If they do, then they count as robust sentimentalists and my argument applies—*mutatis mutandis*—to their position as well. Prinz's theory is thus not especially liable to the objection I will raise; it is simply the only theory whose *explicit* resources are sufficiently rich for it to be possible to engage with it in terms of the WKM problem. In principle, however, my criticism applies to all theories which share robust sentimentalism's core commitments as specified above. It is thus more for reasons of convenience than for intrinsically philosophical ones that my discussion will mostly be restricted to Prinz's version of robust sentimentalism. In particular, I wish to avoid the impression that I am criticizing the whole family of emotionist theories of moral judgment by criticizing one member—it is actually the strongest member of this family I shall focus on, although this fact is obscured by the failure, on part of the other members, to make the aforementioned crucial distinction.

A few other things need to be noted as well. First, it is important to bear in mind that emotionism is compatible with, but does *not* entail a semantic thesis. It is not (necessarily) a thesis about the workings of natural language as far as it employs moral concepts. As such, it does not entail a commitment to expressivism. And fortunately so, because despite its many followers, expressivism seems to be one of the worst supported theories in the history of philosophy (Schroeder 2008).

Second, it should be noted that emotionism does entail a form of noncognitivism, at least when combined with a noncognitivist account of the emotions. (I will return to this point below.) Its psychological claim says that when people make moral judgments, they do not express ordinary beliefs about what is or is not the case. Rather, they are in an emotional state of mind, and making a moral judgment essentially involves feelings of (dis)approval. Dispassionate moral judgments—that is, sincere moral judgments that are not accompanied by an emotion at a given time—are possible, because a great amount of the content of one's moral views can be committed to memory, and can be retrieved from there without actually having to feel the associated emotion. This dispositional view of the emotions allows emotionists to defend their view against the most obvious objection that very often, moral judges do not actually have an online experience of an emotional reaction while passing their verdicts.

Third, it is important to see that none of the above two theses involves a commitment to the other. One can endorse the psychological thesis that moral judgments consist in emotions and be a realist about moral values. On that account, emotions are simply the epistemic route we have to those values. Moral properties could be part of the fabric of the world, while emotions could be the way we pick up on them. The possibility of this "reverse theory" is often overlooked (Kahane 2012). One can also be an emotionist about the ontology of moral properties and maintain that moral judgments are ordinary assertions about objective states of affairs. This position leads to Mackie-style error theories—if moral judgments purport to be about worldly facts, yet there are no such facts because moral properties metaphysically depend on subjective feelings, then all our moral judgments are necessarily false (Mackie 1977).

2 The Infallibility Problem

Let us now return to our account of emotionism from above. I have said that it comprises a psychological and a metaphysical claim: moral judgments as well as moral properties, the emotionist holds, are constituted by emotions. In what follows, I will argue that this theory makes it difficult to understand how error in the moral domain is possible. Moreover, I will argue that the resources the theory has for showing that moral error is indeed possible are scarce.

Now the charge that emotionist theories of moral judgment—metaethical *emotivism* à la Ayer (1952) and Stevenson (1937)—have problems dealing with the possibility of error is not new. But modern sentimentalism faces this problem for entirely different reasons. In the case of emotivism, the problem was that it seemed to render moral mistakes impossible because it held that moral judgments are not even genuine assertions to begin with. They are not in the business of describing the world, and thus cannot be false. But because they do not aim at truth, they also cannot be true. My argument is different. I maintain that sentimentalism has a problem with moral error not because it says that moral judgments cannot be *false or true*; it says that moral judgments cannot be false because they are *necessarily true*.

Let me explain. On the basis of the definition from above, it is easy to see why emotionism has a problem with error. Many forms of judgments—about the ordinary, medium-sized things that inhabit our everyday world—track the facts of the matter. Under normal conditions, my judgment that there is orange juice in the fridge tracks the fact that there is orange juice in the fridge. I believe that there is orange juice in the fridge because this is actually the case, and I wouldn't believe that it is the case if it weren't. This is because my perceptual beliefs are typically caused by the very states of affairs they are about. Thus, my beliefs can be said to "track" the truth—which is not to say that they are infallible, because there are many ways in which I can be, and indeed have been, wrong about the presence of orange juice in my fridge. I can be in error about this because my memory is fooling me, or because I mistook mango juice for orange juice, or what have you. Now take a look at emotionism: PT and MT taken together make the possibility of error difficult to understand because the tracking relation between moral judgments and moral facts is too close, for on the emotionist account, the state of mind that constitutes my moral judgment also brings into existence the very things it purports to refer to. Jesse Prinz describes the situation like this: "If moral judgments are sentimental, and they refer to response-dependent properties, then the judgment that killing is wrong is self-justifying because killing elicits the negative sentiment expressed by that judgment and having the power to elicit such negative sentiments is constitutive of being wrong" (Prinz 2006, 37).

To state the problem more clearly, let me add some refinements to the theory. For reasons of simplicity, I will focus on judgments of wrongness and the property of moral wrongness. The metaphysical claim says something about what it is to be morally wrong, the psychological claim says something about what it means to judge something to be wrong. Emotionists say that both things have to be understood by way of emotional response. What does that mean? That moral properties are constituted by emotions means that these properties are response dependent: they consist in emotional states of approval and disapproval. Being morally wrong, then, is to be disapproved of, or to cause emotions of disapproval. And to judge that something is wrong is to disapprove of it. If we add these qualifications to the metaphysical and the psychological theses proposed above, and keep in mind that (a) the objects of ethical evaluation are typically actions and that (b) metaethicists typically focus on individual subjects, we get:

Emotionism′

PT′. A subject S judges that an action A is morally wrong iff S disapproves of A.

MT′. An action A is morally wrong iff a subject S disapproves of A.

This account has two explanatory fruits: first, it can explain the automatic and intuitive phenomenology of moral judgment. Sentimentalism

> explains the intuitive character of basic values because the emotions that constitute those values can be known non-inferentially; killing just feels wrong. Moral judgments are self-justifying because the emotions that we experience when we grasp those judgments are also responsible for making the judgments true: moral facts are consequences of our emotional reactions. (Prinz 2007, 88)

Moral judgments consist in emotions, and moral facts are *consequences* of those emotions.

Second, it can neatly accommodate the internalist intuition that there is a tight conceptual link between moral judgment and motivation. According to the emotionist, this link is due to the fact that moral judgments are, in and of themselves, motivational states, namely emotions. In making moral judgments, we apply moral concepts, and a

> concept is a representation of a property. Concepts represent by being reliably or lawfully caused by property instances. Moral properties are powers to cause emotions in us. How do we mentally represent such properties? The obvious answer is that we represent them emotionally. … Moral concepts incorporate the emotions that are caused by moral properties, and thereby serve as reliable detectors for those properties. Emotions are motivating. … Moral judgments dispose us to act because they contain moral concepts. (Prinz 2007, 89)

But both of these explanatory fruits come at a high price, as the above quotes make clear. This price is infallibility and, resulting from it, the WKM problem. The reason why the sentimentalist can harvest those fruits is that she has couched both the metaphysics of value and the psychology of value judgments in terms of emotional reactions. But because of this move, it has become unclear how one's moral judgments could ever become separated from the moral properties these judgments are about: it has become unclear how one's moral judgments could ever be false. If the facts that make our moral judgments true come into existence as a result of the fact that we have made those judgments, then we must always be right. Error is impossible.

3 Recalcitrant and Flimsy Feelings

Why is the claim that our moral judgments are infallible so implausible? Because it is subject to a form of bootstrapping problem: it seems to generate moral truths and moral properties out of nothing, simply in virtue of the attitudes one happens to have, however outlandish and arbitrary they might be.

Now one could simply accept the infallibility problem as a bullet the emotionist has to bite. Indeed, emotionists think it is actually the case that our moral attitudes are arbitrary, and valid only relative to individuals, culture, or society. In order to drive the above point home, we need independent reason to reject the (implausible) strategy of biting the bullet of infallibility.

And there is such independent reason to eschew this strategy, because infallibility makes genuine changes of mind hard to understand. Emotionist theories of moral judgment have massive problems with changes of mind and moral learning, because they have difficulties accounting for the role of two different kinds of feelings to which I will refer as *recalcitrant* and *flimsy*, respectively. Consider the following example: I have been brought up in a homophobic environment, but have come to know better, and no longer endorse attitudes of disgust toward homosexual acts. I have changed my mind about the moral status of homosexuality. But if my feelings are recalcitrant (Brady 2008), and thus just won't go away, then—on the emotionist account—I still *do* believe that homosexuality is wrong, because the emotion is the judgment (Prinz 2006). Moreover, it still is true that homosexuality is wrong (for me), because my feelings constitute the property. Should we then say that I now hold contradictory moral judgments? This seems absurd (Alfano 2009). And should we say, even more absurdly, that both of my contradictory judgments are true?

It seems that the sentimentalist does have to say exactly that, and therefore, I haven't changed my mind about the wrongness of homosexual acts in the first place. Rather, while my original moral belief—embodied in my recalcitrant feeling of disapproval toward homosexuality—still lingers on beneath the surface, I have acquired a new, opposing belief that homosexuality is not wrong, a belief which is constituted by a feeling of approval toward it that peacefully coexists with my original aversive attitude.

But, as I wish to insist, I *have* changed my mind about the issue. Quite simply, the problem is that the feelings I have been brought up to believe have proven to be recalcitrant. I do not endorse them anymore. The emotionist cannot account for this case. That is why biting the bullet, and accepting the infallibility of moral judgment as a theoretical cost worth bearing, is such a bad strategy.

Emotionism is inadequate when it comes to handling changes of mind in yet another way. The example just given shows the emotionist's inability to deal with cases in which a subject changes her mind about a moral issue while some of her feelings of disapproval remain. Many cases of genuine changes of mind about moral questions exist that emotionism cannot classify as such. But furthermore, there are cases in which subjects clearly do *not* change their mind in which emotionists seem committed to saying that they do. Prinz cites the phenomenon of "repeated exposure" (Millgram 1999) here. If I am repeatedly exposed to something that used to cause certain very strong emotional reactions in me, these emotions will weaken over time and ultimately disappear. Prinz writes: "repeated exposure to, say, people who are homeless will lead to a reduction in our feelings of outrage, but ... we do not say that homelessness is less wrong than it was when we first encountered a homeless person for the first time" (2007, 97). Here is Prinz's telling solution to the problem of flimsy feelings: "As a first line of response, I am tempted to bite the bullet here and say that over-exposure ... can alter our sentiments and thereby alter our views about what's right and wrong" (97–98).

There are two problems with this. First, it seems wildly implausible that, upon walking around the corner and seeing a second homeless person the sight of which does not stir up the same strong feelings of outrage, I have changed my mind about the matter within seconds. Why should I have? Second, and perhaps more mysteriously, why would the moral wrongness of the fact that there are homeless people change as a function of the feelings I happen to have right here and now? Not only have my flimsy feelings made me change my mind; the moral facts themselves have become flimsy.

Now Prinz thinks that emotionism can escape this problem, because repeated exposure does not change the *dispositions* to experience certain emotions that I have committed to long term memory. This is of course true, and it thus seems that emotionism can account for the required judgmental and factual continuity. But eventually, repeated exposure will have changed my emotional dispositions as well, and I have changed my moral mind when all that happened, is that I have been numbed by the grave injustices I cannot help but witness daily. It is of course still possible for me to retrieve from memory the—meanwhile entirely dispassionate—judgment "It is so wrong that there are all these homeless people." But either way, this judgment is not accompanied by any feelings whatsoever anymore. Then emotionism, which is committed to the modally strong claim that there *cannot* be moral judgments without at least some emotional component, is falsified. Or, I am not making a genuine moral judgment anymore, because my belief has been deprived of all feeling. I am merely reporting, inverted comma style, what genuine, emotionally involved moralizers think. All of these options are of course much less plausible than the simple truth: that I have *not* changed my mind about homelessness in an instant, and that my emotional state of mind makes no essential contribution to the question whether I have done so or not.

4 The Wrong Kind of Mistake

There are further reasons why one might want to be able to account for the possibility of moral mistakes. One of them is that once one has denied the possibility of moral error, it becomes increasingly difficult to make sense of our "critical practices"—the practices of justifying and criticizing our own moral judgments and those made by others. In fact, it becomes utterly mysterious why there are these practices in the first place, because they seem to presuppose the possibility of error. Mark Timmons writes: "The adequacy of a metaethical view is judged in part by how well it comports with common-sense moral practice and, in particular, with how well it comports with our critical practices concerning moral judgment, including the presumption that all sorts of error in moral judgment are possible" (1999, 84). Here is a list of possible sources of moral error we are given by Timmons (1999, 83–84):

(i) error in nonmoral belief;
(ii) error rooted in nonmoral reasoning;
(iii) error with regard to various constraints on moral reasoning;
(iv) error because one's moral judgment does not conform with one of the following (or both):
  (iv.i) the moral convictions one would have (under counterfactual ideal conditions)
  (iv.ii) the moral norms and values one's community would have (under counterfactual ideal conditions).

I have already said that a theory of moral judgment, whether emotionist or not, owes us a threefold account. It ought to have something to say about the *attitudes* moral judgments consist in, the *objects* these judgments are about, and the *information* that is relevant to how the two are connected. Take cognitivist realism about moral judgments and values as an example. Cognitivists say that moral judgments are beliefs: they purport to be about moral states of affairs. This is the *attitudinal* part. Realists say that the *objects* of moral judgments are moral states of affairs. If these states of affairs obtain, they are moral facts. These facts are real: they exist objectively and (in some sense) mind independently, and are part of the fabric of the world, very much like stones, cars and H_2O. The *information* that is relevant to how moral beliefs are connected to moral facts—how we find out what the moral facts are, so that we can represent them correctly with our beliefs—is straightforward information about what is right or wrong, or what one has most (moral) reason to do. Moral errors occur whenever we believe something to be right/wrong that is in fact not right/wrong. I do not endorse cognitivist realism here. But it is a good example of a theory that has no problems dealing with moral error.

Now take emotionism. Emotionists hold that the attitudes that constitute moral judgments are emotions. The objects these judgments are about—moral properties—are also emotions. And the information that is relevant to how the two are connected is plain factual information. There are thus three ways in which moralizers, according to emotionism, can make moral mistakes. They can be wrong about what they approve or disapprove of. They can be wrong about the facts. And they can emotionally respond to the wrong objects. I shall discuss each of these options in turn, and argue that none of them captures the *right kind of mistake.*

(1) Being wrong about what one approves of

In this section, I will argue that emotionism conflates moral error with moral insincerity. It says that people make moral mistakes not because they do not judge truly, but because they do not judge truthfully. This is the wrong kind of mistake.

According to emotionism as described above, one can publicly announce to be caring about the poor, and to think that social inequalities are deeply wrong, and meanwhile harbor no such feelings. The kind of mistake involved in this is not one of saying something false, of making a moral error: either, one is lying (because lying requires an intention to say something false, not that what one is saying is in fact false), or one is deceiving oneself, and is getting one's own feelings wrong. One can have mistaken descriptive beliefs about what nondescriptive attitudes one has. This is such a case.

Why is being wrong about how one feels not the kind of mistake we are thinking about when we are thinking about moral mistakes? Why is it the wrong kind of mistake? Suppose you are making the judgment that it is deeply wrong—though not necessarily imprudent, as superstition would have us believe—to walk under a ladder. People who walk under ladders make you shudder, they disgust and outrage you. What would it be like to be wrong about what one approves of? To be wrong about one's own feelings? Being wrong about what one feels is to believe that one has certain feelings or emotions, yet it is not in fact true that one has them. It seems to me that this phenomenon is not as common as psychoanalytic theory would have us believe (knowing *what* one feels is different from knowing *why* one feels the way one does). If walking under ladders is disgusting to you, you will probably know it. Genuine moral mistakes, however, occur very often. They consist in being mistaken about what is valuable, which moral norms are acceptable, and what virtuous conduct would have been. We constantly misjudge what is the right thing to do, and have to struggle to get it right as well as we can. If moral mistakes consist in being mistaken about one's own feelings, the pervasiveness of moral error would be left unexplained.

Still, being wrong about what one approves of is not entirely unusual: "Show me a young Conservative and I'll show you someone with no heart. Show me an old Liberal and I'll show you someone with no brains." This transition happens, I suppose, piece by piece, and slowly creeps up on most people who undergo it. During this transitional phase, there must be many points at which one thinks of oneself as a liberal, but does not in fact share typically liberal sentiments. But there are more straightforward cases as well. People often suddenly realize that they do not love their partners anymore, and haven't loved them for quite a while. Or they come to learn that they do not really like going to the opera, but simply thought they did, or pretended they did to please their opera-loving friends.

How do emotionists deal with cases like this? Here is how Prinz describes some of the experimental findings that bear on this issue:

> They [Wheatley and Haidt] found that some people who were hypnotized to feel disgust ended up morally condemning a perfectly innocent individual. I think such condemnations qualify as errors because they were not caused by sentiments in long-term memory, but rather by extraneous facts; they do not qualify as legitimate expressions of the subjects' moral attitudes. In short, a wrong action is an action against which an observer has a moral rule. If an action is condemned because of hypnotically induced disgust, it does not qualify as wrong. (Prinz 2007, 96)

Why, according to Prinz's sentimentalism, do people's moral judgments in the emotion induction experiments qualify as errors? The reason for this is not that they judge something to be wrong that is not really wrong. They are making a moral mistake because their judgments were not "legitimate expressions" of their "moral attitudes." Hypnotically induced disgust tampers with people's ability to accurately introspect their feelings. They are tricked into thinking they are disgusted by something that they do not really find disgusting. Subjects are wrong about the matter at hand because they do not really have an emotion-backed moral rule against the acts they are made to condemn. They are inadvertently insincere.

But this is not what we have asked for. We were looking for an account of moral error, an account of what it is to make moral mistakes. That is, we wanted to know how to frame the relation between moral attitudes and moral properties such that erroneous relations of that kind become intelligible. What we have been given, however, is an account of the relation between the higher order descriptive beliefs subjects have about their moral emotions, and how *those* beliefs can be mistaken.

The subjects in the experiments make mistakes because they make incorrect moral judgments. What would happen upon a postexperimental debriefing? How would subjects react if the hidden source of their disgust were revealed to them after the fact? I suspect that they would do two things, rather than one: they would be surprised that the scientists were able to pull off this stunt and trick them. That is, they would be surprised that it was possible for them to make such a grave mistake about what they "really" approve of. And this is the important part: they would give up their false judgment. This means that they would correct both: the introspective mistake they made about their feelings, and the moral mistake they made about the case at hand. This shows that the two are not the same. And that the emotionist conflates the latter with the former.

Now consider the contrast case in which a genuine moral error is revealed to someone. For the sake of the argument, suppose that it is morally wrong to eat meat. Now suppose that there is a morally decent carnivorous man who does not think it is wrong to eat meat; this man knows where meat comes from and how it is produced, and that it typically requires animals to be bred in species-inappropriate conditions and be killed. One day, our decent carnivore stumbles upon video footage online of how this breeding and killing actually works, and becomes a vegetarian. This man has made a moral error—he thought it was morally acceptable to eat meat, but in fact it is not. But this error was not due to false factual information, as he knew what there was to know about the meat industry; it was not based on a misrepresentation of his own feelings, either. He did not, for all we know, have an opinion about how he felt toward animal cruelty, which then became falsified by the reality of his emotions. Rather, he made a value judgment that turned out to be false. Now it is of course true that it was his emotional reaction that made him realize that eating meat is wrong.[5] But this reaction merely constitutes the man's revised judgment. For the example to work, one need not assume that the property of wrongness his response is directed toward consists in the emotion as well.

(2) Being wrong about the facts

The infallibility problem highlights the fact that emotionism's psychological and metaphysical theses establish a tracking relation between subjects' moral judgments and the moral facts these judgments are about that is so close that it renders error impossible. Moral judgments are *necessarily true*, because they create the values they aim to be judgments of.[6] There is, however, one strategy the emotionist can choose in order to loosen the tracking relation between PT′ and MT′. This strategy is idealization.

Here is how idealization works. On the one hand, PT′ holds that moral judgments are states of emotional (dis)approval; but on a suitably amended version of MT′, moral properties are constituted by the emotions one *would* have under suitably specified, ideal conditions. Thus we get emotionism″:

Emotionism″

PT′. A subject S judges that an action A is morally wrong iff S disapproves of A.

MT″. An action A is morally wrong iff a subject S would disapprove of it under conditions of full relevant information and rationality.

This loosens the tracking relation between sentimentalism's psychological and its metaphysical component, because now we can imagine cases where the former becomes disconnected from the latter. I might have endorsed the invasion in Iraq as right because of my belief that Iraq has weapons of mass destruction. I was wrong, however, because the property of being right is constituted by the feelings I *would* have had under ideal conditions. But under ideal conditions, I would have known that there were no weapons of mass destruction in Iraq, and thus the corresponding response-dependent property would not have been instantiated under those conditions. My judgment would have been erroneous, because I would have judged some action to be right (simply by approving of it) which is not really right (because the action does not instantiate the "idealized" property of rightness).

Interestingly, this is the only kind of idealization that works for the emotionist. One could, it wrongly seems, idealize both elements of the sentimentalist account. Call this *emotionism‴:

***Emotionism‴**

PT″. A subject S judges that an action A is morally wrong iff S would disapprove of it under conditions of full relevant information and rationality.
MT″. An action A is morally wrong iff a subject S would disapprove of it under conditions of full relevant information and rationality.

But this clearly does not help, because it merely passes the buck of the infallibility problem onto a higher level. Now I still cannot be wrong, because the moral attitude and the property remain on a par with each other.

One could also, it seems, idealize the account the other way round, and loosen the tight connection between emotionism's attitudinal and its metaphysical element by idealizing PT′ rather than MT′ resulting in *emotionism⁗:

***Emotionism⁗**

PT″. A subject S judges that an action A is morally wrong iff S would disapprove of A under conditions of full relevant information and rationality.
MT′. An action A is morally wrong iff a subject S disapproves of it.

Here, the judgment is understood in terms of what response one would have under ideal conditions, while the property is understood in terms of what response one just happens to have. But this account delivers some utterly strange results. In this proposal, our moral judgments always seem to be ahead of the properties, as it were, and there are cases in which my informed judgment would have to be classified as being false, due to the fact that it did not correspond to a property that is constituted by my uninformed attitudes. Let me illustrate this with an analogy. Take perceptual judgments, and stipulate that when it comes to visual perception, conditions of plain daylight are ideal. Now let's say that perceptual judgments are constituted by the perceptual experiences I would have under ideal conditions (of plain daylight). And let's also say that, by analogy to *emotionism'''', colors, which are response-dependent properties, are constituted by the perceptual experiences I happen to have under actual, possibly nonideal, conditions. Now suppose that I am looking at my bookshelf in somewhat darkish lighting conditions. All my books appear to be gray. According to the proposed account, this makes them in fact gray. Now if I were put in ideal conditions of plain daylight, my books would not all appear to be gray. This constitutes my perceptual judgment. But it would also mean that my judgment about what colors my books have would be in error, because I did not get the facts right, namely that they are all gray. Similarly, if moral *properties* were constituted by *actual* states of (dis) approval, and moral *judgments* were constituted by *ideal* states of (dis) approval, more and better information could lead me to making erroneous moral judgments about properties that are constituted by less and worse information, provided that everything else is equal. That is an unhappy result indeed, and it is not entirely clear which option emotionists actually endorse. Prinz writes:

> If "wrong" referred to whatever causes disapprobation in me, then I could not judge something to be wrong in error. To avoid this consequence, we must idealize. We should say that the word "wrong" refers only to those things that irk me under conditions of full factual knowledge and reflection, and freedom from emotional biases that I myself would deem as unrelated to the matter at hand. (Prinz 2006, 35)

On one reading of this passage, idealization affects the judgment; on the other, it affects the property, on another, it affects both.

It seems fair to criticize only the strongest possible version of the theory. As we have seen, only emotionism″ helps tackle the infallibility problem. *Emotionism‴ does not solve the problem. And the "solution" offered by *emotionism'''' is not a real solution either, because it has absurd implications. But emotionism″, although it is the most promising form, also makes the biggest problem for this account to come to the fore. Are there any genuinely *moral* mistakes, or are there only *factual* mistakes about moral issues? Some have tried to reduce moral mistakes to mistakes about the factual basis of our moral beliefs (Stevenson 1966). If we assume that the moral belief "Abortion is wrong" is incorrect, this incorrectness can be reduced to a combination of a presumably correct moral belief (or an expression of a feeling which is not truth-apt, respectively) and an incorrect factual belief (*Killing persons is wrong/Boo to killing persons! + Fetuses are persons*). Emotionism" can deal with false factual information. For instance, I have made a judgment about the moral status of abortion by disapproving of it. But my judgment is incorrect, because due to the fact that moral properties consist in the feelings I would have under ideal conditions, and that under ideal conditions I would know that fetuses are not persons (because they do not possess the capacities required for this status), my judgment does not correspond to the moral facts.

But first, even with this in hand we do not know what *moral* errors consist in. Since we have separated a distinctively ethical from a distinctively factual component in our moral judgments, we still lack an account of what it would mean to make a moral mistake that is not due to factual misinformation. Second, it has become questionable whether in the proposed account, there could ever be such a thing. If we can always offer a description of my faulty moral judgment in terms of some factual error that fueled my emotional reaction, then my moral mistakes will always, and necessarily so, boil down to mistakes in my descriptive beliefs. This is the wrong kind of mistake.

(3) Being wrong about the objects of one's judgments

Being wrong about the attitudes or the information one has does not give the emotionist an account of the right kind of mistake. Let us see if the third and final option fares any better.

If cognitivism about the emotions (see, for example, Nussbaum 2001) is true, then the emotionist position—at least to the extent that it is supposed to be a rival for rationalist accounts of moral judgment—loses its bite. If emotions simply are cognitive states, then the challenge toward rationalism does not cut any ice.

For this reason, emotionists about moral judgment are often noncognitivists about the emotions. That emotions are noncognitive states does not mean, of course, that they do not represent anything. Indeed, they are often taken to represent their "formal objects": sadness represents loss, fear represents danger, disgust represents the threat of contamination. This seems to give the emotionist the desired account of moral error. If emotions represent something, then they will do so in error if they represent it incorrectly, or if they represent the wrong objects.

Two questions need to be addressed here. First, how do emotions represent? And second, how is emotional misrepresentation possible? As for the first question, the noncognitivist will want to maintain that emotions represent nonconceptually. Judgmental states are neither necessary nor sufficient for the experience of an emotion (Scarantino 2010). Rather, emotional intentionality consists in the causal link they have to the objects that reliably cause them. The concept "horse" represents horses because it is reliably caused by horses. That it is sometimes caused by cows does not matter here, because it is not *reliably* caused by them. Pleasure represents (physical) well-being, because it is reliably caused by it. In fact, that is the function it has been selected for: it has been "set up to be set off" by states of well-being (Prinz 2007, 61).

Moral emotions must have been set up to be set off by something as well. That is how they represent, and it is also what gives us the standards that determine whether an emotion represents correctly. If it picks up on what its function of detecting is, it represents accurately. Take guilt as an example. Some have hypothesized that guilt "represent[s] the concern expressed by: I have violated an autonomy rule against a member of a group with which I feel a connection" (Prinz 2007, 76). If this is supposed to give us an account of genuine moral error, then it should comport well with paradigmatic cases of morally wrong actions, and thus be extensionally adequate. Paradigmatic cases of morally wrong actions should be among the things that our moral emotions are set up to be set off by. I think that the naturalist account of what it means to respond to the wrong objects with one's emotions cannot meet this requirement.

That the naturalist account of the moral emotions delivers extensionally inadequate results is nicely illustrated by the first horn of Sharon Street's "Darwinian dilemma" for moral theories (Street 2006): if some things are morally wrong, and if emotional responses pick up on the things which are morally wrong, then these responses are correct. But if our emotional responses are also shaped by natural selection, then we have no reason to believe that these emotions pick up on the moral facts, that is, that they are correct. In fact, we have reason to doubt that they do, because the pressure of natural selection drives us toward what will help us survive, not what is morally right. Presumably, our moral emotions have been set up to be set off by what has survival value, rather than moral value.

Of course, the emotionist could reject the presuppositions this dilemma makes, and maintain that what is morally right is determined not by moral facts, but by subjective emotional responses. Thus the fact that our evolutionarily shaped emotional responses do not track any moral facts need not bother the emotionist. But this reply, rather than help him out, directly brings the emotionist back to the infallibility objection I have raised above. If there is nothing left to track for our emotional responses but those emotional responses themselves, then the resulting account will either commit the naturalistic fallacy—because it is now forced to hold that what renders an emotional response correct is that it is reliably caused by the factors it has evolved to pick up on, which is to say that the moral collapses into the advantageous—or face a recoil into infallibility.

This is a substantial flaw, and it already makes an affirmative answer to the second question superfluous. What use is an account of the right *kind* of mistake, if it cannot also give us an account of the right *set* of mistakes? The problem with the naturalistic account of what emotions represent is thus not, strictly speaking, that it does not give us the right kind of mistake, but that its naturalistic commitments drive it toward an implausible evolutionary background story. And saying, as one might, that something's survival value is constitutive of that something's being morally right is a fallacious last resort that will not work, either.

One could say, of course, that moral mistakes consist in emotional disapproval toward things which are not really morally wrong. That sounds about right. But this proposal hardly sounds emotionist anymore, because it is incompatible with its metaphysical thesis about the ontological status of moral properties, as these were said to be the "consequences of our emotional reactions" (Prinz 2007, 88). A recourse to quasi-objective moral facts is thus not an option for emotionists.

What, then, are genuine moral mistakes? I will answer this question very briefly, by citing some negative characteristics of moral errors and a few of fairly straightforward examples. Moral errors are errors in moral judgment which are not due to factual or introspective errors. Let me give you a few of examples of genuine moral mistakes (see Huemer 2005, 102). Some concrete ones are:

(1) Killing children in order to eat them is right.
(2) Torturing people for fun is good.
(3) One ought to take advantage of others as often as one can.

Some slightly more abstract ones are:

(4) Pain is better than pleasure.
(5) It is right to punish a person for a crime he or she did not commit.
(6) If X is better than Y, and Y is better than Z, then Z is better than X.[7]

All of these are genuine moral errors, yet none of them can (necessarily) be reduced to errors in the factual basis of one's moral beliefs or errors in beliefs about what one (dis)approves of. If emotionism is true, then we are either infallible moral judges or incapable of making the right kind of mistake.

5 Substantive Moral Mistakes

Emotionists cannot account for genuine moral error. But do they want to do so? And, more importantly, do they need to? Advocates of the emotionist theory of moral judgment could point out that in asking for an account of genuine moral error, I am begging the question against their position. It seems that I am demanding from the emotionist that she explain something she thinks does not exist. In saying that there must be such a thing as *genuinely moral* error—mistakes that go beyond error that is due to false factual information or faulty introspection—I am making a presupposition the other party might not be willing to share. This leads to an unhelpful standoff.

It is of course very hard, if not impossible, to motivate someone to make a conceptual distinction she does not want to make. I suggest that there is a meaningful distinction between *genuine moral error* and *nonmoral error in the moral domain*. But one could simply ignore that distinction. I wish to argue that emotionism did not keep its promise to deliver an account of real moral error. But then again, emotionists will want to insist that this is a misunderstanding, and that they have not really made *that* promise at all: they have acknowledged that there is some room for error in moral judgment, but hold that it does not go further than the kinds of nonmoral error the theory can easily accommodate. However, not making a conceptual distinction comes with the cost of losing conceptual sophistication. To my knowledge, there is no emotionist *explanation* for why most people are making a distinction that is entirely unjustified, and no emotionist *justification* for why we should stop making this distinction.

The only way to avoid this unhelpful standoff, it seems to me, is to show that emotionists do in fact share the presupposition at issue here, and do make a distinction between moral error and error in the moral domain that comes from a different source. In fact, Prinz himself, in his rejection of the necessity of empathy for moral judgment, has recently argued that empathy is often "epistemically unreliable" (2011b, 223). If there are no standards of correctness, this assertion makes no sense whatsoever. Still, the standards of correctness Prinz is talking about here could be nonmoral in nature. Let me explain why they are not.

Why is empathy supposed to be an unreliable epistemic guide for our moral judgments? Empathy can lead us astray because "our capacity to experience vicarious emotions varies as a function of such factors as social proximity and salience. … If you are a stranger or if you are located in a distant land, my degree of empathy may be correspondingly reduced" (2011b, 223). This is arguably due to evolutionary causes, as my dispositions to feel empathetic have developed in my phylogenetic past, which did not give my ancestors (who lived in communities of a rather small scale) the opportunity to evolve a disposition to experience contagious distress in response to the suffering of distant others.

Obviously, this argument owes a lot to Peter Singer's (1972, 2005) idea that our moral responses to salient suffering and suffering that we are not confronted with as vividly are inconsistent. The question, then, boils down to whether this inconsistency, and thus the error it gives rise to, is of a distinctively moral kind. If it is, then we have shown that emotionists presuppose a substantive notion of moral error; if it is not, then the kind of inconsistency we are talking about here is merely logical, and could be grouped among the familiar types of nongenuine moral error.

What grounds do we have to think that the above inconsistency in our responses is due to morally insubstantive logical reasons? Think about Singer's "shallow pond" example. We think that it is obligatory to save a drowning child from a shallow pond even though it might ruin our new suit, but, although we think that it would be morally praiseworthy to do so, we do not think that we have an obligation to save a child that is starving to death in a foreign country by making a similarly insignificant sacrifice. One could argue that this inconsistency is merely logical: we endorse as a moral rule the principle that harm and suffering are bad and should be prevented, combine it with a rule of proper moral deliberation that says that we ought to treat like cases alike, feed these two principles with the description of the two cases, wait for the output—our moral judgment—and find that, if the respective judgmental outputs differ, we are making a mistake. Our judgments are inconsistent.

But this cannot be right, because we cannot detect this inconsistency on mere logical grounds. Rather, "the inconsistency arises out of our responses to the two cases in the context of our basic substantive moral commitments that inform us about what is morally relevant" (Campbell and Kumar 2012, 296). Note that people's differential responses to the "base case" (the proximal drowning child) and the "target case" (the distal starving child) are not inconsistent because people are responding differently to *identical* cases, but because they are responding differently to a target case that is different from the base case in many ways, though not in any *morally relevant way*. What is and is not morally relevant, however, is not picked out by the harm principle; it is not picked out by the rule to treat like cases alike (because that rule can only be applied once we know whether two cases are alike in morally relevant respects); and it is certainly not picked out by the factual information we have about the two cases. It is picked out by our substantive moral commitments, among which we find that salience and spatial proximity do not matter for moral evaluation.

And, as Campbell and Kumar have convincingly argued, these substantive moral commitments are located in our emotional System I, which is the mental subsystem that emotionists think is responsible for moral judgment: "First, we identify the salient differences between the two cases; second, these differences are fed as input to system [I]; third, if none of the norms in system [I] are activated—if the difference is not perceived as morally relevant, thereby engaging system [I]—system [II] issues in a negative affective response" (2012, 291). Perhaps it would be better to describe this as an affectively neutral response with respect to the question whether the second case is morally different, which then results in an affectively negative second order response to the perceived inconsistency in one's judgments. But irrespective of that—if people are making a moral mistake here, as Prinz seems to suggest when talking about "epistemic unreliability," then this mistake is substantively moral: people judge that we have no obligation to save the starving child, but they are wrong about this. Thus, we have a form of genuine moral error here, and evidence that emotionists are committed to its existence.

Conclusion

This chapter and the next two belong together. I announced in the introduction to this book that I am rejecting the claim that emotion and reason are incompatible with each other, or that the two concepts even lie on the same categorical level. This is the reason why the emotional basis of moral cognition, the evidence for which I will discuss later, does not make me uncomfortable. At the same time, this makes it legitimate to ask why, given the significance of emotions for moral judgment, I avoid referring to my account as a version of sentimentalism. This chapter was the first part of my two-part response to this question.

I have argued that emotionism, understood as a comprehensive metaethical doctrine that comprises both a psychological and a metaphysical thesis about the nature of morality, cannot satisfyingly deal with genuine moral error because it eschews making strong normative commitments. In the next chapter, I will show that there are sentimentalist theories which are not as normatively prude; I will provide reasons to reject those theories as well, which concludes my response to why I prefer to call the educated intuitions model rationalist.

5 Are Emotions Necessary for Moral Judgment?

Introduction

I am walking along Amsterdam's Grachtengordel, the city's world-famous belt of canals. I notice the splendid patrician mansions, and stop by at one of the cozy "brown cafés." I am intrigued by the florilegium of small shops offering antique furniture, vintage clothing, and other curiosities until I accidentally pass by the Torture Museum; I take a quick look at the poster on the museum's front door, and whatever I pick up about the chairs and masks, cages and forks, needles and blades that were used as means for torment and inquisition makes me think: "This is just wrong." But—what am I thinking here? What state of mind am I in? How did I arrive at my judgment? And, not least, is it justified?

Traditionally, philosophical metaethics was thought to offer two different answers to these questions—rationalism and sentimentalism. Here is how one might caricature these two positions: Rationalism claims that, in thinking "Torture is wrong," I am thinking that torture is wrong; that I, and any other person, ought not to do it and that there is good reason for this; that it cannot be willed without contradiction to torture people or that it is simply immoral and, hence, irrational. It claims that I am in a cognitive state of mind, comparable to judgments about the most efficient means to cause the strongest pain in an unlucky suspect, yet thoroughly *sui generis*. And it claims that I have arrived at my judgment through careful weighing of reasons and conscious reflection, and that I have adopted the one attitude toward torture that finally survived my incorruptible scrutiny. Sentimentalism, on the other hand, claims that I am in a state of emotional arousal, that I have arrived at my judgment unconsciously, through an emotionally triggered disgust response toward torture and the contagious distress that the images of excruciated bodies have caused in me; and that my judgment "Torture is wrong" is supposed to spread the word and make other people feel the same kind of disapproval toward this atavistic practice.

There must be some middle ground between those caricatures. Rationalism is strongest in highlighting the distinctive role of reason, reflection, and self-criticism. Without these, we wouldn't understand our practice of moral judgment anymore (and wouldn't like it, either). Sentimentalism seems to be an attractive position when it comes to explaining the emotional dimension associated with moral judgment: we care about our values, very deeply indeed, and we experience anger or guilt upon the violation of the moral norms we endorse. How could it not be true that the realm of moral values and norms is not the object of cool and sober contemplation, but of passionate engagement and emotional commitment? Sentimentalism tries to embrace this connection between emotional reactions and moral judgments.

Empirical moral psychology has taken up the challenge of providing such a middle ground. It tries to leave the empty space of empirically frictionless conceptual analysis and to offer a scientifically credible account of the relation between emotional reactions, conscious reasoning, and moral judgment. Experimental philosophers (Knobe and Nichols 2008; Appiah 2008) have paid a lot of attention to empirical findings about moral judgment and agency, claiming that philosophical metaethics must be pursued in a descriptively adequate fashion. Moreover, experimental philosophers claim that normative ethics does not tell us anything informative and interesting about flesh-and-blood human beings if it is based on a stark opposition between "is" and "ought." If "ought" implies "can," then "cannot" implies "not ought." From this, it is only a small step to "if never happens, then probably cannot, then probably ought not." A moral theory that tells us what people ought to do, regardless of whether *anybody* has *ever* done it, or will ever do it, must eventually fail. (This reflects a commitment to Flanagan's "principle of minimal psychological realism" [1991, 32].) Or so empirically minded philosophers argue.

Naturalistic accounts of moral judgment and reasoning—like the previously discussed SI model—promise to offer clear and conclusive answers in the long-standing debate between rationalism and sentimentalism about moral judgment. They use respectable methods and publicly verifiable procedures, particularly highly sophisticated social psychological and neuroscientific experiments. For obvious reasons, there has been a trend to draw broadly sentimentalist conclusions from the available evidence: the philosophical account of moral judgment that empirical moral psychology seems to favor is the view that emotional and intuitive "gut reactions" rather than genuine moral reasoning are crucial for normative judgment. As Daniel Jacobson puts it, "Social intuitionism, considered as a thesis of moral psychology, best coheres with a *sentimentalist* metaethical theory, which holds that (many) evaluative concepts must be understood by way of human emotional response" (Jacobson 2008, 220).

According to the anti-rationalist challenge, moral judgments are based on intuitive gut reactions; moral reasoning is a means for social persuasion. I have argued that this challenge fails: there is nothing dubious about the intuitive character of moral judgment, because reason *becomes* intuitive over time. Similarly, there is nothing problematic about the social nature of moral reasoning, because we are better at reasoning in packs than we are at deliberating on our own.

Despite the elective affinity, the emotionist challenge is independent of all this. No matter what role reasoning plays for moral judgment, the values we reason about might still be grounded in feelings. Now as a conceptual point, this is most certainly true. And I am happy to agree that values *are* based on emotions. I have also remarked, however, that this fact does not challenge rationalism about moral judgment, because emotion and reason need not be considered incompatible.

In what follows, I will briefly review and assess the experimental evidence in favor of the claim that emotion is essential for moral cognition. My reading of this claim is that emotions are both *necessary* and *sufficient* for moral judgment. I shall develop an interpretation of both these claims that I deem to be satisfactory for a moderate rationalist position like the one I envision here. I argue that this type of rationalist about moral judgment can be happy to accept the necessity thesis. My argument draws on the idea that emotions play the same role for moral judgment that perceptions play for ordinary judgments about the external world. Provided an empirically adequate and normatively convincing interpretation is available, the same holds for the sufficiency thesis. I develop such an interpretation and show that it can successfully account for the available empirical evidence. The general idea is that the rationalist can accept the claim that emotional reactions are sufficient for moral judgment just in case a subject's emotional reaction picks up on the morally relevant features of the situation. This is all the moral rationalist could ever ask for. Now the experimental evidence suggests that sometimes, this does not happen. But when it doesn't, I will argue, no genuine moral judgment is made, so that that the empirical evidence does not undermine the rationalist's position.

This is the main line of my argument. Toward the end of the next chapter, I will attempt to explain, in a purely metaethical and normatively noncommittal fashion, some of the basic requirements for a response to count as a genuine moral judgment. My proposal is that an emotional reaction amounts to a genuine moral judgment just in case the response toward an action in question causes the judgment in a reasons-responsive way that can be reflectively endorsed under ideal conditions. This is but one attempt to cash out what it takes for an emotion to pick up on something of moral relevance. This should, however, leave my main point untouched: that there are certain constraints on the concept of a moral judgment that make it immune, at least to some extent, to the most radical forms of empirical debunking. Now what empirical research could show is that there is no such thing that satisfies those constraints, and thus no such thing as moral judgment. But to my knowledge, no one has made this latter claim, nor does there seem good reason to do so.

I will start with the necessity thesis, the discussion of which has four sections. In section 1, I explain how the thesis is borne out empirically. Section 2 spells out the perceptual analogy. The third section addresses some important reservations people have about empathy. Finally, section 4 asks whether the very evidence for the necessity thesis concerning *emotion* and morality does not also show that *reason* is necessary for moral judgment.

1 The Necessity Thesis: Psychopathy and the Moral/Conventional Distinction

Moral judgments are related to norms, and norms are related to human action. These norms are prescriptive: they specify what *ought* to happen or what *ought* to be done. Examples for prescriptive norms are the rules of the road as well as norms of etiquette and, in general, rules that govern the interactive space between persons. But prescriptive norms can be quite different in nature, depending in part on the source of their authority. Take, for example, the norm to shake a person's right hand upon meeting her. And now take the norm not to torture people out of boredom. Both norms are prescriptive. But intuitively it seems that the validity of the first norm depends upon a mere convention. We can say that *you ought to shake a person's right hand upon meeting her*. If the norm were, however, to shake with the *left* hand, it would not be wrong to shake with the right. It is after all a mere convention. Emotionists about morality claim that in order to fully understand the fundamental difference between conventional and nonconventional norms, one must be able to experience certain emotional reactions toward their transgression. They claim that in order to grasp the specific moral authority of certain norms, one must be susceptible to experiencing guilt or outrage upon their violation. More generally, being susceptible to feeling an emotional reaction toward certain types of norm transgressions is seen as a psychologically necessary feature, an enabling condition for moral judgment. Call this the *necessity thesis*.

There is overwhelming agreement among philosophers and psychologists about the fact that for a person to be able to make moral judgments she must be able—among other things—to draw a distinction between moral and conventional norms (Nichols 2004, 3; Nucci 1985; for criticism of this claim, see Sinnott-Armstrong and Wheatley 2012). On Shaun Nichols's "sentimental rules" account, for example, the capacity for "core moral judgment" is introduced along the very lines of the moral/conventional distinction; moral judgment requires the capacity to understand a certain subclass of prescriptive social rules as nonconventional, transgressions of these norms as more serious, generalizably wrong (that is, wrong in other countries or communities as well) and the validity of these rules as neither based on social acceptability nor dependent on authority. All these criteria spell out the nonconventional validity of moral norms. Note that it is notoriously tricky to cash out the exact details of the moral/conventional distinction; in what follows, I will mostly rely on an intuitive familiarity with the distinction. The most sophisticated recent account of the distinction has it that conventional norms require that their justification—the rationale for why one ought to do as they say—make an essential reference to an established social practice of acting according to the purportedly conventional rule (Southwood 2011; Southwood and Eriksson 2011). Moral norms, by contrast, do not essentially make such a reference. For the purposes of the discussion below, however, the details of this will not matter.

Ever since the days of Phineas Gage,[1] research on psychopathy and so-called acquired sociopathy has provided the best and most robust evidence for the thesis that a certain kind of emotional engagement is necessary for moral judgment and behavior (Saver and Damasio 1991; Damasio 1994). In psychopaths, we find two things combined: first, a highly impaired emotional makeup, and second, a reliably poor performance on tasks to draw the moral/conventional distinction (see the classical research conducted by Turiel 1983 and Blair 1995; for more recent discussions of the phenomenon, see Hare 1999 and Blair, Mitchell, and Blair 2005). Many empirical moral psychologists claim that the former is an enabling condition of the latter. In order to show that psychopaths are unable to draw the moral/conventional distinction, James Blair predicted that psychopathic patients will have difficulties not only in drawing the distinction, but that this incapacity will also show on the justificatory level: their justification for why a certain norm transgression is "wrong" will be "less likely to make references to the pain or discomfort of victims than the non-psychopath controls" (Blair 1995, 13). Another prediction was that psychopaths were likely to treat moral rules as conventional rules. Surprisingly, the latter prediction turned out to be false. Subjects were presented with several stories (a child hitting another child in the moral case, a child talking in class in the conventional case) they had to assess in light of the question whether the described transgressive behavior should be seen as a violation of a moral or a conventional norm. Blair found that psychopaths treated all transgressions as moral and the validity of the transgressed rules as authority independent. But this doesn't, according to Blair, show that psychopaths fail to make the moral/conventional distinction because of an *increased* moral sensitivity, or an increased tendency to empathize with the victims of moral transgressions. One has to bear in mind that all the test subjects (psychopaths as well as nonpsychopaths) were inmates (most of them convicted murderers); they simply had a strong motivation to demonstrate that they had improved, or even acquired new social and moral knowledge, and this made them overshoot the target. Psychopaths do fail to grasp this important feature of genuine moral judgment, only in an unexpected way. And lying to the interviewers in this way is exactly the kind of thing a real psychopath would do.

The most natural explanation for this seems to be that psychopaths lack the capacity to "feel" the special character of violations that are wrong, regardless of what an authority would think about them. This specifically moral sensitivity is developed at an early stage and is remarkably robust (see Nucci 1985, who found that children from the Amish community treat moral rules as independent even from God's authority). In combination with the evidence about the overly formal ("You are just not supposed to do it!") justifications psychopaths offer for their judgments and their shallow and undifferentiated emotional life (Blair, Mitchell, and Blair 2005), the evidence suggests an important link between moral judgments and affective capacities. Moreover, acquired sociopaths and patients whose sociomoral emotions are impaired due to damage of the ventromedial prefrontal cortex (VMPFC) are less likely to respond to highly emotionally engaging moral dilemmas—such as the "smothering the baby" case—in the same manner as normal subjects (Koenigs et al. 2007; Kennett and Fine 2008, 173–174). However, it would be premature to conclude at this point, as Jonathan Haidt does, that the "very existence of the psychopath illustrates Hume's statement that ''tis not contrary to reason to prefer the destruction of the whole world to the scratching of my finger'" (Haidt 2001, 824). Psychopaths might simply have lost the ability to respond adequately to the requirements of reason.

In a slightly different context, Adina Roskies (2003) has argued that it is both empirically and conceptually implausible to assume that people who suffer from so-called acquired sociopathy are unable to make genuine moral judgments. She argues that there is no evidence that, as their judgments remain adequate, the moral knowledge of patients with focal damage to the VMPFC is impaired as a result of their condition, that it would be implausible to argue that the content of their judgments has changed or that their injury has turned them into detached observers who merely report other people's moral beliefs in an "square quotes" sense. This is an empirical counterexample which seems to cast the necessity thesis in general into doubt. If the phenomenon of acquired sociopathy shows that proper motivation is not entailed by the capacity to judge morally, then the motivational deficits psychopaths exhibit do not establish any form of modally strong necessity of emotion, either.

Let me note that Roskies's arguments do not undermine, but might actually strengthen the rationalist's dialectical position. If she is correct, then emotions might not even be necessary for moral judgment, and the challenge the necessity thesis allegedly poses for the rationalist disappears. (I am arguing that emotions are not incompatible with reason, so I can remain agnostic about this point.) It might be that psychopaths and VMPFC patients (she only talks about the latter) do make genuine moral judgments, but simply fail to act accordingly (Cima, Tonnaer, and Hauser 2010), which would entail that we are only dealing with evidence for a motivational problem, rather than one that affects subjects' capacity to make moral judgments. I doubt, however, that her argument can achieve this. First, it only applies to cases of adult-onset brain damage. Emotions, and more specifically empathy, might still be *developmentally* necessary for the capacity of moral judgment. Second, it remains true that psychopaths do not fully grasp the moral/conventional distinction. Her argument does not undermine this point. Given that the distinction is one essential element of moral judgment, psychopaths fail to understand one essential element of moral judgment, and hence fail to make full-blown moral judgments.

The emotionist could argue that psychopaths make *abnormal* moral judgments, and that abnormal moral judgments are still moral judgments. This is a legitimate objection, and obviously, there is no clear cutoff point at which an abnormal practice turns into a completely different one. But if we take a look at the various morally important capacities psychopathic individuals lack, including

- a stable disposition for morally inappropriate behavior,
- poor or absent moral emotions,
- failure to grasp the inferential structure of normative judgments (see Kennett and Fine 2008 for examples of so-called retractor statements), and
- failure to grasp the moral/conventional distinction,

it is most plausible to deny them moral competence altogether. When it comes to deciding whether a person's judgment is genuinely moral, one cannot look at singular instances but has to take into consideration a subject's overall status as a diachronic moral agent (Gerrans and Kennett 2010; Damm 2010). Moral judgments are made by morally competent subjects. Psychopaths and (some) subjects with acquired sociopathy do not pass this test. The phenomenon of psychopathy does suggest that emotional responsiveness is necessary for moral judgment and agency.

2 Perceptual Characteristics of Emotions

Emotionists draw on the observation that at least beyond the realm of abstract rights and duties, agents need to be emotionally involved in order to really grasp the complex infrastructure of everyday morality. In many cases, rational insight does not suffice, especially when it comes to motivation: there is something dubious and even callous about a person who visits his friend out of a sense of duty, and not because of his feelings of friendship. (Accordingly, emotionists often endorse a version of the "one thought too many" argument; Williams 1981) Moral psychologists and experimental philosophers drive the claim that emotions are necessary for many important aspects of morality even further, arguing that the very capacity to make moral judgments and act on them fundamentally depends on a person's emotions. I have called this the necessity thesis and discussed the evidence from psychopathy and acquired sociopathy that is cited to support it. Psychopaths and acquired sociopaths suffer from severe emotional impairments, which makes them insensitive to the "practical oomph" that is typically connected to moral imperatives (Joyce 2007). But, does the rationalist have to worry about the claim that emotional reactions are necessary for (or lack thereof damaging to) moral judgment? Let's put things into perspective.

Is the necessity thesis plausible? On a certain reading, the answer is "yes." Emotions complement moral reasoning and contribute to the phenomenal richness of moral experience; but this doesn't necessarily downgrade moral reasoning to an ineffective epiphenomenon. The right way to conceive of the significance of emotions for morality, I suggest, is along the lines of a modified version of Kant's famous statement: moral thoughts without emotional content are empty, moral emotions without reasoning are blind. Just as perceptual content is causally necessary to provide our judgments with objective material and prevent them from a "frictionless spinning in a void" (McDowell 1994), emotional reactions are causally necessary to provide our moral judgments with normative content. And just like perceptual appearances—think about the Müller-Lyer illusion—give defeasible reasons for belief, emotional responses give defeasible reasons for normative judgments. In the light of reasons to distrust our emotions, however, we have to rely on cognitively more elaborate principles of practical reasoning.

Some rationalists about moral judgment have succumbed to the temptation to argue for "pure" reason in the realm of morality. The evidence for the necessity thesis suggests that this is not a good idea. Rather, they should be happy to accept the emotional *impurity* of moral judgment and endorse it as an integral part of the moral life of human beings. Otherwise, rationalists will have to face what might be called the "pure-reason objection," an objection that has been raised by Michael Slote, Jesse Prinz, and others: "Sure, if rationalism is true, we don't need the sentiments; we can rely on our rational cognition" (Slote 2004, 13). But can we even conceive of morality as having a "purely cognitive source" (Prinz 2006, 33)? And if morality is supposed to make efficacious demands on us—motivate us to act—can it be based on nothing but theoretical cognition? (This is one aspect of what Smith [1994] calls the "moral problem.") "Pure" rationalists about morality and externalists about motivation hold that it can, and maybe they're right. At any rate, I shall try to meet the emotionist about morality half way and pursue the more parsimonious strategy here. We can at least grant that emotions do play a central role for moral judgment and behavior. So I will argue, at least.

One can, I have suggested, think about the emotional underpinnings of moral judgments in a way that is analogous to the relation between perceptual content and judgments about the empirical world (Goldie 2004b, 2004c, 2007). Advocates of genuine moral reasoning can adopt the view that emotional reactions are necessary to provide content to moral judgments, content that is related to us as feeling and acting human beings. In acknowledging the significance of emotional reactions for our moral thinking, we can even make sense of the fact that our emotional experiences seem to be beyond our rational control: in having emotions, we are passive. From that, however, it does not follow that emotions bear no rational connection to the web of our moral beliefs. Compare the case of perception again. Perception too has an irreducibly passive element. But this only demarcates the line between objective thinking, which stands in front of the tribunal of experience, and fiction, which does not. Passivity is not a threat to the autonomy of a thinking subject, but a necessary condition for objectivity: the fact that our thoughts depend upon the way the world is, a way that is not at our disposal and independent of our arbitrariness. I take emotions and perceptions to share several characteristic features:[2]

(i) Emotions have an *objective phenomenology*: they are experienced as being representations of something outside of them. This holds for simple feelings as well as complex emotions. A pain in the funny bone is experienced as something that refers to a condition of the funny bone, not merely the experiencing subject. Guilt or remorse, on the other hand, are experienced as representing something blameworthy in the agent deserving of guilt or remorse. Emotions are not experienced as combinations of facts and an emotional resonance in a feeling observer that is entirely disconnected from the events the emotion is directed toward. Bear in mind that this is a phenomenological, not a metaphysical point. An objective phenomenology need not reflect a corresponding relation between emotions and objective moral facts.

(ii) In having emotions, we are *passive*. It is often not up to us which emotions to have. To be subject to an emotion is something that can merely happen to a person. It typically does not require cognitive effort (like imagination or concentration) to have them, persons do not have to find out inferentially whether they are in a state of emotional arousal or not and, usually, what kind of feeling or emotion they are subject to at a particular moment (pain, grief, and so on). This does not mean, of course, that we have no control whatsoever over which emotions we experience. We can choose which emotion-eliciting situations to expose ourselves to, or work actively toward better emotion management, for example, through psychotherapy.

(iii) Although emotions are often beyond our rational control, they are *rationally amenable* to some extent. This feature might also be called the reason-responsiveness of emotions. They are the proper object of evaluation and critical reflection, and we can—and often do—ask whether having a particular emotional response to a particular emotionally significant event makes sense. This is sometimes called the "Rational Assessment Problem" (Prinz 2007, 60). The rational amenability of emotion does not entail, however, that emotions themselves are inferentially structured. Nevertheless, they stand in the space of reasons: it makes sense to give and ask for reasons for emotions.

(iv) Emotions can have an *informative nature*. In particular, they can inform a person about new values in unexpected ways, very much in the same way as perception can trigger new and unexpected beliefs:

> Emotions are like perceptions in that they can arise independently of our considered convictions about the circumstances eliciting them, and may even conflict with those convictions.

And,

> when they persist despite those opinions they induce us to question the opinions. So even if I think beauty is only skin deep, and being smart or interesting or funny is what's important, I can be brought to think that one's appearance matters more than I previously acknowledged by finding myself ashamed of my flabby stomach at the beach. (D'Arms 2005, 9)

The informative nature of emotional experience has also been borne out empirically, in terms of the so-called "affect as information" paradigm (Schwartz and Clore 1983).

(v) Emotions are, just like perceptions, phenomenally *fine-grained*. As feelings, they have a level of subjective detail that cannot be fully exhausted by conceptual means alone (Gunther 2003). Novelists like Proust have worked hard to capture in words how humans experience emotions, but their exact phenomenal character eludes exhaustive description.

(vi) Just like perceptions, emotions transcend themselves toward an *intentional object*. People feel guilty *about something*, they resent others *for something*, and they are afraid *that something might happen*. This special kind of intentionality has been described as a "feeling toward" (cf. Goldie 2000; Döring 2007). Kenny's (1963) distinction between an emotion's material and its formal object is instrumental here: emotional intentionality is directed both toward a particular object, such as the anger one feels toward yesterday's insult by a colleague, and toward a formal object, such as offenses in general. And while this might not apply to all emotions, it does apply to those emotions which are most central to the study of moral judgment, such as disgust, empathy, or guilt.

(vii) Emotions and perceptions essentially involve a *perspective* that other mental states such as beliefs, for example, lack (de Sousa 1987; Deonna 2006). My belief that the table is round and your belief that the table is round are the same belief with the same propositional content. But my perception of the table and your perception of the table, as well as my and your experience of the concert last night, are irreducibly different due to their difference in perspective. You stood in the first row, I stood

> somewhere in the back. You positioned yourself to the left of the stage, I ended up somewhere to the right. These differences bring with it certain perspectival differences that are characteristic of sense experience. Similarly, the way the concert resonated emotionally with me will be different from the way it affected you, and this analogous perspectival difference is reflected by how our overall emotional experience unfolded over time.

I cannot develop a full-blown theory of the emotions here, but let me briefly discuss the main problems there are with the emotion/perception analogy. On the face of it, the oscillation of emotions between their passivity and rational amenability seems difficult to explain. That may be, but it only makes the emotion/perception analogy more plausible. In fact, this ambivalent relation to the space of reasons seems to be one of the key issues in current epistemology of perception (McDowell 1994). Peter Goldie has suggested that in order to reconcile the elements of causal passivity and spontaneous activity of emotions, one can draw a distinction between two kinds of inferentiality: "a belief can be both non-inferential in the phenomenological sense (no conscious reasoning by the subject) and yet inferential in the epistemic sense (justifiable by the subject)" (2007, 349). Emotional experiences typically occur noninferentially, but are susceptible to reasoning after they have emerged.

Moreover, the emotion/perception analogy is the best way to deal with the problem of recalcitrant emotions (cf. Benbaji 2013). If emotions are, as the James Lange Theory has it, mere feelings (i.e. they consist in the conscious awareness of patterned bodily changes), then it remains puzzling how there can be experiences of emotional recalcitrance in the first place. If emotions are, as cognitivists argue, primarily judgments of value, then the phenomenon of emotional recalcitrance consists in a conflict between unconsciously and consciously held evaluative judgments (Brady 2008). This is implausible (Scarantino 2010). Emotions that just won't go away are best described as analogous to optical illusions.

But aren't emotions, one might object, unlike perceptions in that they are not directed at the external world, but at an inner realm of subjective experiences that is only accessible through introspection? No, they aren't, because emotions are essentially connected to bodily changes: they are, among other things, feelings that detect such changes. But the body is part of the world. Perceiving what happens with and inside the body is perceiving a part of the world, the only difference being that this very part of the world is *me* (or any other perceiving subject, respectively).

There is one crucial difference between emotions and normal perceptions one has to bear in mind: perceptions tell us how the world is. Emotions, as affective-motivational states, tell us how the world—according to the person who has them—ought to be. Thus emotions cannot be said to refer to facts of any kind, nor that they are subjective experiences of evaluative facts (as McDowell [1985] seems to suggest). They don't represent the world, but an agent's position in the world, and suggest a range of possible goals to pursue or states of affairs to avoid.

Now one might think that this normative direction of fit poses a problem to the grounding of moral norms: if emotions concern what ought to happen rather than what does happen, then they must lack the authority and credibility of scientific beliefs. And this is of course true—but first, it should be noted that it is hard to see how this could be otherwise, given that moral judgments guide action rather than beliefs; and second, it is difficult to spell out this worry in a way that does not beg the question: the emotionist challenge has it that moral judgments are not based on reason because they require emotions. My response was that emotion and reason need not be seen as antagonists, and that the idea behind this challenge, namely that affective-motivational states are epistemically problematic, does not add up. Now according to the above objection, one worry I have not yet been able to eliminate was that emotions have a normative direction of fit. But this worry does not go beyond the challenge originally put forward, which said that affective-motivational states are somehow problematic—and this is exactly the point the emotion/perception analogy casts doubt on.

In the case of theoretical cognition, epistemic agents struggle for knowledge about the world. Their immediate raw material is sensory perception. In the case of practical reason, moral agents try to figure out what to do, and their immediate raw material is emotion. A good reason for a belief about the empirical world is one that is based on sensory perceptions everybody could share; analogously, a good reason for action—or, third-personally, judgment about action—is one that is based on motivational states (desires, emotions, or sentiments) everybody could share. On that view, emotions fit nicely into a normative perspective on moral judgment. Perceptions present the world to be in a certain way, a way that is subject to error and illusion. They nevertheless present a way the world appears to be—they are "is-appearances," as it were. Emotions figure in moral judgments in the same way as perceptions figure in perceptual judgments: they are, as it were, "ought-appearances" and defeasibly present a way the world ought to be according to someone's individual perspective. Consider the following, final illustration: if you witness an old lady being mugged by a young man you will respond to that event with distress, a feeling that is connected to a disposition to experience empathy upon the harm being done to other people. In this case, your empathy alerts you as the witness that something fishy is going on, something that calls for further attention or action. If it wasn't for your emotional reaction, you might well have overlooked the incident altogether. Our emotional reactions help us to stay *open* to new moral experiences and present to us the fine-grainedness of the moral world; they are the material that forces us to recalibrate our moral convictions, the driving energy in a perpetual process of moral change, and a force of stimulation, rather than deformation, for moral thinking (Arpaly 2003).

I have argued that emotions are part and parcel of human practical rationality. I have also argued that the reason for this is that emotions play the same role for moral judgment and decision making that perception plays for judgments about everyday objects and events. But, as Karen Jones (2006) has recently argued, these two claims are in tension with each other, because perception is supposed to be *modular*. One feature of mental modules that is particularly unwelcome from the perspective of the educated intuitions account is the fact that perceptual mechanisms are informationally encapsulated. The necessity thesis fits into this account because of the way in which moral intuitions are emotionally charged. But if these emotions are like perceptions, then they are impervious to higher cognition; however, it is one of the main claims of the educated intuitions account that emotionally charged moral intuitions are *not* impervious to reasoning. This is a problem.

The *locus classicus* for an account of modularity is Jerry Fodor (1983). Modular mental processes are said to share various characteristics. Fodor mentions nine: specific neural localization, automaticity, high speed, specific breakdowns, inaccessibility, simple outputs, domain specificity, ontogenetic determinacy, and informational encapsulation.

As far as the educated intuitions account is concerned, informational encapsulation poses the most pressing problem. I want to counter the modularity problem that looms for any account that combines the emotion/perception analogy with the idea of an education of the emotions on two grounds. First, there is the empirical observation that examples for genuine informational encapsulation are surprisingly hard to come by. One often used example is the recalcitrance of perceptual illusions mentioned above. Regardless of the information a subject has about the actual length of the two lines in the Müller-Lyer illusion, the faulty perception does not go away. This is because the perceptual system is modular.

But as Joseph Henrich and his colleagues (2010) have observed, there is evidence suggesting that cognitive impenetrability comes in degrees, and that even the modularity theorist's favorite examples are susceptible to a surprisingly large amount of culturally induced variation. In short, there are studies suggesting that the Müller-Lyer illusion is a cultural artifact created by growing up in "carpentered" societies in which people are surrounded, from very early on, by straight lines and right angles. The San foragers of the Kalahari, for instance, are not subject to this illusion. But if even the most basic purportedly modular cognitive systems such as perception are not thoroughly encapsulated, what is?

Moreover, and this is my second reason for rejecting the idea that the emotion/perception analogy and the notion of an education of the emotions do not go well together, there is positive evidence to think that even the most primitive feelings and emotional responses can be "recalibrated" (Prinz 2004, 99). Emotions reflect their evolutionary history by being set up to be set off by certain things: they are calibrated to pick up on some triggers and not others. But through learning, they can be recalibrated; over the course of this process, experiences and judgments can attune our emotions to respond to new triggers, such that "we can come to be afraid of electric shocks and the erosion of civil liberties under anti-terrorist laws" (Jones 2006, 20). Recalibration frees our emotions from their evolutionary history, at least to a certain extent. It is here where their "wisdom" lies, rather than in some obscure sense of repugnance whose authority we must accept on good faith (cf. Kass 1997 for this proposal).

One of the main virtues of the emotion/perception analogy is that it nicely captures the way different emotions pick up on different evaluative domains, very much like sense organs are dedicated to the detection of specific stimuli. Disgust—more on which later—picks up on violations of purity; indignation concerns transgressions of fairness; empathy is about harm to others; and so on.

The success of the strategy of defending rationalism against the emotionist challenge on the basis of the emotion/perception analogy hinges on two things: first, on how plausible one takes the analogy itself to be. In this section, I have tried to lend some credibility to it. Second, it depends on how forceful one takes arguments with the above structure to be. The perceptualist response to the emotionist challenge employs a "companions in guilt" strategy (Mackie 1977, 39). Such a strategy takes one allegedly problematic set of claims whose acceptability it tries to vindicate; then, in a second step, it takes another, allegedly *un*problematic set of claims and points out that this second set shares the very features which are supposed to render the first set problematic. This strategy is especially popular in metaethics (Lillehammer 2007), where it has been used to defend the objectivity of value, for instance, by comparing it to scientific practice. What I have done in this section was to point out that emotions, which the emotionist thinks cannot provide a rational basis for moral judgment, share many of their purportedly problematic features with perceptions, which are typically not thought to be problematic in the same way. They are passive, modular, perspectival, and outstrip our conceptual capacities; but none of these characteristics should be considered threatening. Indeed, they should not even feel all that surprising.

3 Morality and Emotion: The Limits of Empathy

Moral rationalists can embrace the fact that emotions are necessary for moral judgment: moral judgments without emotions are empty, moral emotions without cognition are blind.

However, this perceptual model, as one might call it, is not the most obvious and straightforward way for the rationalist about moral judgment to incorporate the necessity thesis. The best way to do this, it seems, would be to argue for a cognitivist position in the theory of emotion. If emotions just are cognitive states (evaluative judgments), then it is even more clear that the fact that emotions are necessary for moral judgment does not have to intimidate the rationalist the slightest bit. (For an overview of this position, see Deigh 1994. For a defense of cognitivism about emotions, see

Solomon 1976; de Sousa 1987; Greenspan 1988; Helm 2001; Stocker and Hegeman 1996; Nussbaum 2001.) I do not have much to say about emotional cognitivism in this book. If it is the correct account of the emotions, so be it. But I doubt that it is, for the simple reason that cognitivists about emotion fail to capture what is emotional about emotions: they over intellectualize them (Goldie 2000). This is not supposed to be an argument against cognitivism. It is not even a sketch of one. It is merely an explanation of why I have not gone down the cognitivist route, and why that is not a problem for me: as far as defending a rationalist position about the psychology of moral judgment goes, the truth or falsity of emotional cognitivism either helps my argument or leaves it unaffected.

Another point I wish to emphasize is that the claim that emotions are necessary for moral judgment has been discussed in terms of whether one particular emotion, namely empathy, is necessary for moral judgment. Empathy makes us sensitive to and concerned about the suffering and well-being of others. But there are of course other aspects of morality, many of which empathy has no bearing on whatsoever. On the CAD model (Rozin et al. 1999), for instance, morality comprises issues regarding the protection of the norms of the community (C), individual autonomy (A), and divinity/purity (D). Only one of those—autonomy and the rights of individuals to do as they please—has a connection to empathy, and an indirect one at that. The morality of the community and of sacredness, as Rozin and his colleagues point out, are accompanied and supported by emotions such as contempt and anger, rather than empathy. Graham, Haidt, and Nosek (2009), whose work I have already mentioned in the introduction, talk about at least five (more recently: six) moral foundations—care/harm, fairness/cheating, loyalty/betrayal, authority/subversion, and sanctity/degradation (more recently: liberty/oppression). Only the first of these five can be accounted for in terms of the necessity of empathy. The others require emotions such as respect, indignation, or disgust for subjects to become susceptible to their moral force. Let it suffice to say here that the rationalist can agree with all of this. There is a wide spectrum of emotions which are necessary for people to become sensitive to the widest spectrum of morally relevant features. I have focused on empathy here. But if my rationalist rejoinder to the necessity thesis works for empathy, then it does so for other emotions as well.

Some emotions are necessary for full-blown moral agency, but in a slightly different way. Guilt and shame, for instance, do not directly ground people's emotional reactions toward moral norms and their transgression. Guilt and shame are moral emotions that sit on top of other moral emotions. People feel guilty when they have done something wrong. But whether something is morally wrong is not directly determined by whether people feel guilty upon doing it. Rather, some actions are wrong because, for instance, they harm others, and people respond to this kind of harm with empathy. Our empathetic response tells us that the suffering ought to stop, or that it should not have been inflicted in the first place. And when we have done the damage already, we can start feeling guilty about it.

This suggests that empathy is not the only thing that is necessary for moral judgment. There are things next to empathy, such as disgust, that matter as well. And there are things on top of empathy, such as guilt, that play an important role, too. Some emotionists, however, argue that emotions in general are constitutive of moral judgment, but that empathy is not part of those morally relevant emotions. Jesse Prinz (2011a, 2011b; see also Decety and Cowell 2014), for instance, joins Kant and Nietzsche with his plea against the moral significance of empathy by developing an elaborate argument against the various ways in which the emotion of empathy might be considered important for moral judgment. He attempts to show that empathy is not necessary for moral judgment either constitutively, causally, developmentally, epistemically, normatively, or motivationally. One can come up with many cases (such as victimless crimes) in which neither *agent-empathy*—empathy with the motives of an agent—nor *patient-empathy*—empathy with those affected by the actions of an agent—constitutively grounds subjects' moral (dis)approbation, or in which their moral reaction is clearly based on some other feeling, such as contempt; for obvious reasons, empathy and vicarious distress are of no importance whatsoever for my moral judgment that someone was wronged when that someone was me; the developmental role of empathy can easily be overestimated, because the developmental differences found in psychopathic individuals can be attributed to their flattened affect in general; empathy as an epistemic guide toward moral judgment is often unreliable, because it is sensitive to morally irrelevant factors such as familiarity and spatial proximity; the motivational pull exerted by empathy is often weak; and, finally, empathic responses play no direct role in the justification of moral judgments: what makes an action wrong is the harm it causes, not the empathy that is caused in a witness.

I take most of Prinz's points here, but would like to insist that his arguments do not show that there is no way in which empathy ever matters to moral judgment. Rather, his arguments serve to show that the role of empathy has been seriously overestimated by large chunks of the philosophical tradition, especially among emotionists. This is probably correct, but it does not undermine the idea that empathy often plays an important epistemic, motivational, and developmental role for people's capacity to make moral assessments of a situation, even though it can easily be manipulated and does not cover all, or even most, of what morality is about. We believe that harming others is wrong. But it is hard to imagine why we should believe this if it were not for the empathic response we experienced upon other people's suffering. On the other hand, people have suffered vicariously not only for the plight of the American slaves, but also for the hardship of their owners who had to do without them, which shows that to produce good moral judgments, empathy requires a good deal of rational guidance.

Moreover, there is further empirical evidence that empathy is not as important for morality as one might intuitively think. Jeanette Kennett, for instance, has argued that the ability of people with autism spectrum disorder to make moral judgments and develop moral agency casts doubt on the emotionist idea that empathy is crucial to moral judgment. Available evidence suggests that autism is associated with diminished empathy, and numbed responses to the distress of others (Frith 1989). In fact, one of the most recent descriptions of autism spectrum disorder explains the condition in terms of an "empathizing-systemizing theory" (Baron-Cohen 2009): subjects suffering from autism show an increased propensity to systemize, as indicated by their need for repetitiveness, attention to detail, and their narrow interests, as well as an impaired capacity to empathize with other people (see, however, Jones et al. 2010, who argue that autists do not suffer from diminished empathy but merely from impairments of the cognitive mindreading abilities that often underpin empathy).

Kennett argues that autists do not typically suffer the kind of "moral death" (Murphy 1972) that can be found in individuals with psychopathic tendencies, which seems to suggest that a lack of empathy need not result in the complete loss of moral agency. Autists do care about morality; they want to do the right thing, whatever it might turn out to be; very often, it is just not obvious to them what that right thing is. Accordingly, Lisa Damm (2010) and Victoria McGeer (2008) argue that empathy deficits cause many autists to have *epistemic* problems with figuring out what the right thing to do is in many situations, whereas psychopaths are incapable of being morally *motivated*. The latter know full well what the right thing to do is (Cima, Tonnaer, and Hauser 2010)—they're just not interested in doing it (unless it happens to serve their interests).

4 Is Reason Necessary for Moral Judgment?

Emotions are necessary for moral judgment. This is one of the key claims emotionists make in order to pose a challenge to rationalism about the psychology of moral judgment. I have shown why the rationalist can accept this thesis, even though the evidence for the specific contribution of empathy for moral cognition remains ambiguous. At this point, it seems useful to ask the converse question: is reason necessary for moral judgment as well? Would this help the rationalist make her case?

The evidence from psychopathy can be used to pass the ball back to the emotionist. Psychopathic patients suffer from characteristic moral deficits. This, it has been suggested, is due to their diminished emotional responsiveness, particularly when it comes to empathetic responses. But psychopaths, as well as so-called acquired sociopaths, suffer from characteristic rational deficits as well. Ignoring this fact seems to be begging the question against rationalists.

I have already mentioned the fact that psychopaths are prone to be inconsistent in their verbal behavior, exhibiting a variety of so-called "retractor statements." Lisa Damm quotes the following passages from Robert Hare's (1999) interviews with psychopathic offenders: "When asked if he experienced remorse over a murder he'd committed, one young inmate told us, 'Yeah, sure, I feel remorse.' Pressed further, he said that he didn't 'feel bad inside about it.' When asked if he had ever committed a violent offense, a man serving time for theft answered, 'No, but I once had to kill someone.'" Here is another one: "Asked how he had begun his career in crime, [a psychopath] said, 'It had to do with my mother, the most beautiful person in the world. She was strong, worked hard to take care of four kids. A beautiful person. I started stealing her jewelry when I was in the fifth grade. You know, I never really knew the bitch—we went our separate ways'" (Damm 2010, 279).

Some people have taken these kinds of statements to show that psychopaths are less sensitive to the phenomenon of cognitive dissonance. They do not feel as strong a need to resolve inconsistencies between their expressed opinions and their behavior, or between contradictory things they say. They do not seem to perceive conflicts within their agency—because there might just be no moral agency within which such conflicts could arise. If there is no thread of a continuous agent running through one's particular beliefs and desires, inconsistency does not matter.

Among the criteria used to diagnose psychopathy are a variety of traits that indicate a lack of key features of practical rationality: their lack of realistic long-term goals and their impulsiveness make them pursue petty projects at the expense of their more important goals; this can become so extreme that it becomes questionable whether they have any values or important goals at all; their grandiose sense of self-worth remains unshaken by evidence; they continue to lie, cheat, and manipulate even when it is clear that trying to convince the people around them and gain their trust has become utterly hopeless. In short: their characteristic traits make psychopathic people behave irrationally, make them irresponsive to salient evidence, and cause them to systematically act against their best interests, even their immoral ones.

This point has been pressed particularly hard by Heidi Maibom, who offers a thorough analysis of the rational shortcomings of psychopathic subjects. Following Onora O'Neill (1998), she offers a list of key capacities one needs in order to be practically rational (Maibom 2005, 241). Among other things, in pursuing a goal, the rational agent needs to

(i) intend to take the necessary and
(ii) sufficient means to his goal,
(iii) take steps to make these necessary/sufficient means available in case they are not,
(iv) make sure that his goal fits into his overall web of goals, and
(v) make sure that the consequences of his action, to the extent that they are foreseeable, do not contradict his original intention.

Note that this list is normatively parsimonious, and does not smuggle in moral competence as a criterion for practical rationality in order to show that reason is necessary for moral judgment. And also note, as Maibom explains, that psychopaths fall short on virtually all of the above requirements of rationality:

> Psychopathic individuals suffer from principled difficulties that non-psychopathic individuals do not. Compared to the normal population, they have impaired practical rationality. They have problems willing the necessary and sufficient means to their ends, making sure that specific intentions are internally coherent and consistent with the underlying intention, and determining whether foreseeable outcomes of their actions are compatible with their ultimate aim. (Maibom 2005, 244–245)

They want to become rich and powerful, but don't pursue any education or career; they want to lie and manipulate, but continue to contradict themselves; they think their abilities outshine everyone else's, but it is readily apparent that they do not. Citing similar evidence, Kennett agrees with Maibom's diagnosis:

> The evidence suggests that psychopaths have at best a weak capacity to stand back from and evaluate their desires, to estimate the consequences of their actions, to eschew immediate rewards in favour of longer term goals, to time order, to resolve conflicts among their desires, to find constitutive solutions. To these rational shortcomings we may add that psychopaths frequently choose grossly disproportionate means to their immediate ends or fail to adopt the necessary means to their proclaimed ends. (Kennett 2006, 76)

Similar observations can be made about acquired sociopaths and people with focal damage to the VMPFC. Walter Freeman describes patients who underwent frontal lobotomy as

> lazy, rude, boisterous, restless and inane. ... [The patient] is relatively unteachable, having lost those social skills that are necessary for living outside an institution. ... If the patient has previously demonstrated antisocial traits such as alcoholism, drug addiction, criminality, avoidance of responsibility, aggressiveness or psychopathic activities, the effect of operation may be to free him from any residual sense of guilt or shame, and thus turn loose upon society an individual whose behavior is intolerable. (Freeman 1950, 207–208; for an overview over the neuroanatomy of morality, see Fumagalli and Priori 2012)

Damasio and his colleagues describe the effects of brain damage inflicted on patient "EVR" by a tumor as follows:

> By age 35, in 1975, EVR was a successful professional, happily married, and the father of two. He led an impeccable social life, and was a role model to younger siblings. In that year, an orbitofrontal meningioma was diagnosed and, in order to achieve its successful surgical resection a bilateral excision of orbital and lower

medial cortices was necessary. EVR's basic intelligence and standard memory were not compromised by the ablation. His performances on standardized IQ and memory tests are uniformly in the superior range [97–99th percentile]. ... He passes all formal neuropsychological probes. (Damasio et al. 1990, 81–82)

And:

Standing in sharp contrast to this pattern of neuropsychological test performance, EVR's social conduct was profoundly affected by his brain injury. Over a brief period of time, he entered disastrous business ventures (one of which led to predictable bankruptcy), and was divorced twice (the second marriage, which was to a prostitute, only lasted 6 months). He has been unable to hold any paying job since the time of the surgery, and his plans for future activity are defective. (82)

Proponents of the emotionist challenge have a choice to make: either, the abnormal and irrational behavior psychopaths and acquired sociopaths exhibit is due to impairments of their capacity to reason—which means that the challenge the necessity thesis is meant to pose collapses, because now their deficits cannot plausibly be attributed to emotional impairments. Or, they really do suffer from affective deficits. But if emotion and reason are distinct, it remains puzzling why their emotional impairments affect their practical rationality, unless their failures to behave in practically rational ways are evidence for the fact that emotion is a genuine part of practical reason.

A final word on psychopathy: for the purposes of this chapter, I have granted that the findings on psychopathic deficits, in particular in moral/conventional tasks, hold up. But this may not be so. A recent study (Aharoni, Sinnott-Armstrong, and Kiehl 2012; see also Levy 2014) found that Blair's original explanation for why psychopaths appear to be unable to distinguish moral from conventional norms—the incentive of criminal offenders to pass as reformed—cannot be maintained. For when psychopaths are provided with the appropriate external scaffolding (such as a forced choice design in which subjects are told upfront that half of the items belong in each category) to remove the need to appear especially morally conscientious, their performance is restored to normal levels.

Another type of evidence which could be brought to bear on the question of what morality is based on comes from studies with nonhuman animals. In one famous experiment, Sarah Brosnan and Frans de Waal (2003) found that brown capuchin monkeys reject, as their interpretation has it, "unfair" distributions of goods. In their study, the monkeys were trained to exchange a small token for food. In the "equality" condition, two monkeys—who sat next to and could see each other—received a piece of cucumber upon handing the token to the experimenter. In the "inequality" condition, one of the monkeys received cucumber, the other received a grape—apparently a reward which is much more desirable for these primordial gourmets. What they found was that the monkeys virtually never rejected the cucumber when that was what both received in exchange for their token. When one received a grape, however, cucumber was rejected, sometimes fiercely, more than 40 percent of the time. This, they hypothesize, is evidence for the evolutionary precursors of a human sense of fairness. The monkeys do not want to put up with this outrageous inequality. Therefore, they'd rather engage in violent protest.

Now if one is willing to grant, for the sake of the argument, that brown capuchin monkeys—or all monkeys, for that matter—lack sophisticated reasoning abilities of the kind which can be found in human beings, but exhibit a sense of moral desert and fairness, then one could argue that reason is not necessary for basic forms of moral judgment. But on a closer look at the evidence, one can readily see that the rationalists need not be too impressed by this line of reasoning.

First, there are some empirical problems with this argument. As Clive Wynne (2004) has argued, one can hardly overestimate the importance of the fact that in a third condition (which Brosnan and de Waal refer to as "food control"), the mere availability of a more highly valued award—as when a grape was merely presented to the monkeys who received the cucumber, rather than given to another monkey—increased the animals' "inequity aversion" to roughly the same degree. But this cannot possibly be seen as a reaction toward unfair inequality: the grape is merely there, and so they are not receiving an unequal reward for handing the token to the experimenter. It is the mere presence of the tastier alternative, it seems, that makes the capuchins turn down the slice of cucumber. This remains a problem for Brosnan and de Waal's more "normative," fairness-based (Brosnan 2006) explanation of the monkey's behavior even upon taking into consideration their fourth condition (dubbed "effort control"). In this condition, the second monkey receives the grape without having to exchange anything for it. Here, rejection rates are highest, which seems to show that their behavior is sensitive to considerations of merit. And although this is suggestive, the fairness-based explanation simply cannot explain away the more important fact that inequity aversion is comparably high in the food control condition.

This point takes us to some of the conceptual issues there are with this research. These issues are not about correctly identifying the causes of the capuchins' behavior, or about which explanation for it is most plausible or parsimonious, but about whether monkeys' reactions have anything to do with what can count as genuine fairness. Inequality is, of course, sometimes unfair, especially when it is not tied to differential performance and desert. But so is equality: as far as I am aware, it has not been tested whether monkeys would reject the cucumber when another monkey were to receive it without having to exchange anything. Moreover, one should bear in mind that the monkeys who rejected the unequal reward were directly affected by this unequal distribution themselves. It can be doubted whether monkeys are disposed to care about fairness vicariously, and to become angry and outraged when they see a third monkey being fobbed off with something of lesser value. Recently, Michael Tomasello and his colleagues (Riedl et al. 2012) found that chimpanzees do not engage in punitive behavior against transgressors when the victim of the violation is a third party rather than themselves. This impartial viewpoint, however, seems crucial for a sense of genuine fairness. And one could drive this point even further: proper judgments about fairness require the ability not only to be sensitive to the unequal distribution of goods, but also to second-guess the principles that govern those distributive patterns to begin with. Human moral judgment is concerned, not only with whether a given distribution is fair, but also with who ought to be in charge of the distribution of goods in the first place, and whether who happens to be in charge reflects power structures and higher order injustices that are beyond comprehension of even the smartest animal. Now the failure to grasp all the ramifications of the concept of fairness might be due to the fact that monkeys and all other nonhuman animals lack in certain complex capacities for higher *nonmoral* cognition, such as perspective taking, extended foresight, and others. But this would just entail what, on a certain reading of this research, the aforementioned studies were supposed to cast doubt on: that reason *is* necessary for moral judgment after all.[3]

Reason is necessary for moral competence in yet another way. For a large part, the sentimentalist strand in recent empirically informed metaethics has either ignored or only selectively deployed the crucial role general executive functioning and inhibitory control play for moral judgment and agency. But a growing body of research on the antisocial effects of alcohol consumption suggests that normative control is an important part of our moral psychology. To a large extent, moral motivation consists in the maintenance of cooperative tendencies in the face of countervailing inclinations. Immoral tendencies are always there, ready to break through; they must be inhibited in order not to express themselves. Alcohol impairs higher order processing. Since it also leads to antisocial behavior on a grand scale, this suggests that prototypically rational functions matter a lot more for moral judgment and motivation than many sentimentalists would allow (Heath and Hardy-Vallée 2015).

Conclusion

In this chapter, I picked up the empirical thread of my argument. Many philosophers and psychologists nowadays think that questions regarding the nature of moral judgment cannot be answered from the armchair. I share this proclivity for a moderately naturalist methodology. But I do not think that empirical research can answer all of the interesting questions there are. Not that anyone ever claimed that it could; but in the midst of all the excitement and enthusiasm about the empirical turn in philosophy, this fact seems worth mentioning.

One of the things that empirical research has nothing to say about is what the requirements of reason are. Science tells us what is and what happens. About what ought to be and to happen, it must remain silent. Science can tell us whether moral judgment necessarily recruits emotions. But whether *not* recruiting emotions is required for a process to be based on reason, empirical studies cannot tell us. This is where philosophical reflection enters the picture.

6 Are Emotions Sufficient for Moral Judgment?

Introduction

Like the anti-rationalist challenge, the emotionist challenge has two parts. It might seem to some that to establish the necessity of emotion for moral judgment is already a considerable achievement for the emotionist. And it is; but it takes a little more to show that emotions are *essential* for moral judgment. In this chapter, I discuss empirical evidence purported to show that emotions are also *sufficient* for moral judgment, and offer a novel, moderately rationalist interpretation of this evidence from the perspective of the educated intuitions account developed earlier.

I have argued in the introduction that apart from some fairly outdated remnants of faculty psychology in our theoretical idiom, we have little to go on when it comes to arguing that emotion and reason exclude each other. The final two chapters of this book offer a constructive argument for this claim.

I will engage with the empirical data straightforwardly. I review the evidence for the sufficiency thesis, and develop an interpretation of the thesis on which it becomes clear that it does not pose a threat to a moderate, empirically feasible form of rationalism about the psychology of moral judgment. That emotions are both necessary and sufficient for moral judgment does not entail that moral judges do not exercise their rational capacities in making moral judgments. In fact, their doing so is constitutive for the judgments they arrive at to count as *genuine* moral judgments.

This chapter has six sections. Section 1 presents the empirical evidence for the sufficiency of emotions for moral judgment. The second section makes the case that the concept of moral judgment is more demanding than the available empirical research suggests. To produce genuine *moral* judgments, emotions must not merely be causally, but also justificatorily sufficient. Section 3 discusses the concept of reflective endorsement as one way to spell out the notion of this justificatory sufficiency. Section 4 elaborates on the idea of constraints on the concept of a moral judgment. The fifth section addresses two objections: one has it that reflective endorsement is superfluous; the other holds that idealization analyses tend to commit a certain type of fallacy. I respond to and assess the merits of both these objections. Since a good deal of the evidence for the sufficiency thesis concerns the influence of disgust on moral judgment, I close with a general discussion of the relationship between morality and disgust in section 6, and argue that, in principle, moderate rationalists about moral judgment have no reason to disavow disgust as a possible input into moral deliberation and action.

1 The Sufficiency Thesis: Morality and Disgust

A strong form of emotionism has to argue for the necessity *and sufficiency* of emotional (dis)approval for moral judgment. How can the sufficiency thesis be put to a test? What has to be shown here is that people's emotional reactions *alone* are sufficient to explain their moral attitudes. A change in their emotional makeup—that is the hypothesis—will result in a change of their moral judgments, either in their content or, at the very least, in their severity. Subjects' moral judgments will vary against emotional changes much more than they will vary against moral reasoning. Call this the *sufficiency thesis*.

I will proceed strategically, and first take the evidence put forward in favor of the sufficiency thesis at face value. My aim is to show that a rationalist account of moral judgment can be maintained even if the emotionists' best case scenario is true. In the final section of this chapter, I will raise some doubts as to whether it is true, and will discuss more recent arguments suggesting that the effect of incidental disgust on moral judgment may be weaker than previously thought.

There is a considerable body of evidence in support of the sufficiency thesis. Valdesolo and DeSteno (2006) found that contextual variations that induce emotional changes can influence moral judgments. In the footbridge dilemma, subjects usually display a deontological hesitation to sacrifice a person's life in order to save five people from certain death (or, more precisely, to use that sacrifice as a means to the end of saving the five). But people are much more likely to judge it permissible to throw the fat man off the footbridge and prevent the trolley from killing the five people on the track after watching an episode of *Saturday Night Live* that cheered them up a little beforehand. Artificially induced mood changes seem to alter people's moral judgments about the permissibility of killing a person in a dilemmatic case. However, this result does not come as a surprise. We all know from experience that a slight change in one's mood can have a not at all slight effect on one's behavior. Moreover, some have argued that the amusingness of some of the sacrificial scenarios used in empirical studies of moral cognition undermines their evidential value, since it prevents subjects from properly engaging with the cases at hand, thereby undermining their external validity (Bauman et al. 2014).

Complex emotions like amusement are not the only ones to have this kind of impact on our evaluations. The sufficiency thesis can be defended on an evolutionarily more fundamental level as well. Schnall and her colleagues (2008) found that an unpleasant odor—created, for instance, by the use of "fart spray" in subjects' surroundings—can have a significant influence on people's moral judgments about marriage, sex, environmental issues, or the media. They observed the same effect when they put people behind a filthy desk and asked them to judge the permissibility of various public policies. In general, people tend to interpret their own bodily changes—from throat clenching to nausea—as a source of information about an issue of moral significance. They trust their emotional responses to a large extent, and take them to provide cues about the moral status of a described action or event.

Perhaps the most striking support for the sufficiency thesis has been found using the method of posthypnotic suggestion. Take the following experiment that was conducted with a group of highly hypnotizable test subjects. In a 2005 study, Wheatley and Haidt (2005) tested the effects of hypnosis-induced disgust on people's moral judgments. They confronted their subjects with several story vignettes (a congressman taking bribes, cousins performing incest, a man eating his dead pet dog) and asked subjects to assess them morally. Through posthypnotic suggestion, subjects had been primed to experience a quick flash of disgust upon hearing an arbitrary trigger word that occurred in the story (such as "often"). What Wheatley and Haidt found was that in the disgust condition, subjects not only judged morally wrong actions (like a politician taking bribes from lobbyists) to be morally worse than they did in the neutral condition. They judged acts that were not at all wrong (and were judged accordingly in the neutral, no-disgust condition) to be morally blameworthy, too. When subjects were asked what exactly is wrong about, for instance, a student council representative trying to organize events that are interesting for both students and teachers, they started to confabulate, providing far-fetched *ad hoc* reasons ("It just seems like he's up to something") that bore no connection whatsoever to the vignette.

Emotions and feelings alone are, just as the sufficiency thesis claims, enough to account for changes in moral judgment. Subjects' emotional reactions, whether manipulated by environment or a hypnotically induced trigger word, are sufficient to explain people's moral judgments. Some researchers have claimed that feelings of disgust are the very essence of morality: electromyographical evidence about similarities in facial motor activity suggests that moral emotions originate in primitive, but highly adaptive response patterns to contamination and disease (Chapman et al. 2009) and cross-cultural research supports the idea that these mechanisms have expanded from food-related to sociomoral matters in general (Haidt et al. 1997).

Emotionism does not content itself with claiming that emotions accompany moral judgment, or that they follow in their wake. Emotionists want to say that moral judgment is *all about* emotion—that emotion is essential for it. The best empirical operationalization of this essentiality is that emotions are necessary and sufficient for moral cognition. I have already discussed the necessity thesis in the previous chapter. It supports emotionism about morality because if one strips away subjects' feelings, their judgmental and behavioral performance is rendered deficient. The evidence for the sufficiency thesis supports emotionism because it shows that in order to manipulate people's moral judgments, one need not manipulate their descriptive moral beliefs—say, by repeating a moral proposition over and over again like a mantra, until they start believing it. To bring about a change in their moral verdicts, it is enough to change their emotional responses. (It should be noted, however, that these cases are the absolute exception, and that none of the subjects in Wheatley and Haidt's study were made to actually *change* their moral judgments from right to wrong or the other way round. Hypnotically induced disgust only seemed to affect the *severity* of their verdicts.)

Take a look at Eskine's (2011) study on how disgust affects morality ratings. When subjects were given bitter and/or disgusting beverages to drink before completing a moral judgment task, they judged the vignettes more harshly. If one leaves everything as it is, changing only the emotional state of the judging subject, her moral evaluation of the given scenario will change, even though nothing of moral relevance in the given scenario was changed.[1] But let me emphasize the importance of this last point: if nothing morally relevant changes, yet people's judgments do, can we still meaningfully classify them as *moral* judgments? The following sections attempt to answer this question.

2 Justificatory Sufficiency

How plausible is the sufficiency thesis? And what exactly is it that the empirical evidence shows? Take the experiments in which a filthy desk, fart spray, or hypnotically induced disgust trigger an emotional response that prompts subjects to make a certain moral judgment. In these cases, subjects' emotional reactions are in fact sufficient for changes in their moral judgment. But they are sufficient in a very specific sense: they are sufficient to explain people's judgments, that is, they are *causally* sufficient. This is not, however, the kind of sufficiency one would expect from an analysis of the concept of moral judgment. Surely one can change a person's ethical evaluation of a situation by hitting her over the head with a club, but this hardly proves that clubs and fractured skulls are part and parcel of the practice of moral judgment.

Moreover, that emotions influence moral judgments is an uncontested fact, a truism. As Karen Jones puts it: "The folk already know that emotions influence moral judgement. It is part of the commonsense of moral epistemology that we must be on the lookout for the potentially distorting influence of emotions on moral judgements" (Jones 2006, 47). By naming the *causally* sufficient conditions for something to alter somebody's judgment, one misses the point of giving a theoretical account of something.

The causal story that is told by empirical moral psychology doesn't rule out cases in which the judgments that are caused by changes in a subject's emotional makeup do not count as proper moral judgment. How do we know that what people are doing really is making a moral judgment? Consider the following example by Karen Jones:

> Moral judgements are distinguished from judgements of mere liking or disliking by being answerable to reasons. We challenge, accept, and reject moral judgements on the basis of reasons. ... If someone were to answer a challenge to their judgement that an act was morally right by citing the fact that it was done on a Tuesday, our first response would be bafflement and, if further questioning did not bring the cited consideration somewhere closer to the cluster of considerations we recognize as morally relevant, ... we would conclude that the person lacked competence with moral concepts. (Jones 2006, 49)

I do not want to overstate this point. What I want to argue is that moral judgments are different from *mere* preferences or likes and dislikes in that they involve at least *some degree of normative abstraction* (Rini 2013). Accordingly, if a subject were to answer a challenge to her moral belief that an act was wrong by saying that while considering the act she smelled an unpleasant odor, we would be surprised, to say the least, and reject this answer as confusing, and ultimately illegitimate. Unlike mere expressions of disgust, moral judgments are held to certain standards of relevance and rationality. Irrelevant situational features such as scents or ambient noise do not bear on the moral wrongness of an action, even on the most straightforward sentimentalist account. "Moral" judgments that are based on influences or considerations extraneous to the moral status of an action are deficient to a certain extent. Emotionists about morality have to account for this fact in a way that rules out such cases.

I have called the position that maintains that emotional reactions are essential for moral judgment *emotionism* about morality. Let me introduce a distinction between two different kinds of emotionism, which I will refer to as *simple* and *sophisticated* emotionism. The latter, I will argue, is compatible with a moderate form of rationalism about moral judgment. The account of moral judgment given by empirically supported emotionism about morality is best described as simple emotionism:

Simple emotionism

Emotions are causally necessary and sufficient for moral judgment.

Obviously, one has to understand the necessity and sufficiency of emotional reactions for moral judgment *dispositionally*. Any workable account of emotionism about moral judgment must allow for cases in which the associated emotional response happens to be absent. A subject can sincerely judge an action to be morally wrong even though at a particular moment, she doesn't feel any kind of resentment or outrage upon the deed. It suffices if the appropriate response used to be associated with the judgment in the past, having created a disposition to have the feeling at any given moment in the future, a disposition that might not be realized in any instance. If we also add the fact that typically, the objects of moral judgments are human actions, and that typically, moral emotions can be characterized as states of approval or disapproval, we get simple emotionism′:

Simple emotionism′

For a person to make a moral judgment, it is causally necessary and sufficient to have a disposition to feel an emotional reaction of (dis)approval towards a particular action.

In a nutshell, this is a more elaborated formulation of emotionism's *psychological thesis* that was discussed in chapter 4. One important thing to note is that sometimes, dispositions do not manifest. In this case, it sounds odd to say that a subject has made any moral judgment whatsoever. Emotionists respond, first, that this does not happen very often. Typically, when people have the conscious thought that some action is wrong or event bad, the respective emotional disposition will manifest as well. Simple emotionism′ is supposed to also deal with cases in which the conscious thought is present, but there is no occurrent emotion to accompany it.[2] This holds for all subsequent refinements of the emotionist's claim.

As we have just seen, this account is not equipped to rule out deviant cases of manipulation, hypnosis or, in general, the influence of irrelevant factors on people's moral judgments, which, as we have also seen, undermines the moral competence of the people who are under the influence of these factors; it doesn't rule out cases where a person's emotional reaction is sufficient in a merely causal, but morally irrelevant and hence not justification-conveying way. Most subjects are disposed to feel an emotional response of disapproval toward an action after drinking a disgusting beverage. But the revolting taste of the beverage does not really bear on the evaluative status of the judged action. What is needed is a condition that is sufficient for moral judgment in a way that does not threaten a moral judgment's standing in the space of reasons and, at the same time, a judge's moral competence as a whole. What has to be captured is the inevitable justificatory connection that a person's emotional reactions bear to her moral attitudes.

3 Moral Judgment and Reasons Responsiveness

When it comes to finding the essential components of this *justificatory sufficiency*, we can draw on the idea that a proper standing in the space of reasons can be acquired by the following criterion: if a subject doesn't withdraw her judgment after being exposed to further information (for instance during a postexperimental debriefing about how she came to make her judgment through posthypnotic suggestion) her judgment will not count as a genuinely moral one, provided that the debriefing revealed to her that in making her judgment, she did not respond to the morally relevant features, or at least primarily those features, of the action in question. To put it a little more abstractly: genuine moral competence is characterized by the feature that if a piece of information is added after a judgment is made in light of which the initial judgment is undermined, the judging subject will suspend her belief or adequately back it up with appropriate further grounding.

To cash out this proposal in more detail, one can use the notion of counterfactual, idealized conditions. A subject is to be counted as morally competent if, were a certain undermining piece of knowledge added to her set of beliefs, she would reconsider her moral judgment and refrain from it when necessary. Were a subject to be informed that in making her moral judgment, she responded to empty pizza boxes on her table and a subtle smell of rotten eggs, a competent moral judge would reconsider her judgment and either give it up or cite appropriate moral reasons that "repair," as it were, the initial status of the judgment. This is tantamount to saying that were that subject to possess full knowledge and flawless reasoning abilities already, she wouldn't judge the way she did in the first place. What counts for a moral judgment to be genuine is the way a moral subject reacts after being exposed to perfect information and equipped with perfect reasoning. Does the subject stick to her judgment and the way she arrived at it, and endorse both upon reflection? Or does she withdraw it, and reconsider her initial belief?

Note that on this view, there is no stark opposition between reasons and causes for judgments. The distinction that is used here is one between undermining and benign, justification-conveying causes. My proposal is to understand justificatory sufficiency in terms of benign causal sufficiency. An emotional reaction is justificatorily sufficient just in case it is causally sufficient in a way that can be reflectively endorsed under conditions of full information and rationality. Think again about perception. The acquisition of perceptual knowledge about the world is to a large extent a causal process. Nevertheless, we do not hesitate to describe this process as conveying the justificatory force that is necessary to render a perceptual belief that, say, the sun is shining, into knowledge.

If we add these qualifications to the emotionist account, we get sophisticated emotionism:

Sophisticated emotionism

It is necessary and sufficient for making a moral judgment to be disposed to an emotional attitude of (dis)approbation towards certain actions that causes the judgment in a way that can be reflectively endorsed by the judging person under ideal conditions of full information and rationality.

The account I am presenting here, and to which I will add further refinements below, draws on a distinction between mere states of revulsion or compassion which could be artificially induced by morally extraneous means and *genuine* moral judgments. First, one could restrict the range of emotional attitudes such that only emotional attitudes of *moral* disapprobation are included. Second, one could restrict the range of actions such that only actions that pertain to nonconventionally valid norms are among the things genuine *moral* judgments can be about. The above account settles for the third option, which is to capture what constitutes genuine moral judgments in terms of what a subject could reflectively endorse. This way, the account can remain psychologically and normatively noncommittal. I have said in the introduction that it would be unwise, and not at all theoretically innocent, to start with a definition of the concept of morality. It is not up to psychological and metaethical theories to *decide* what counts as morally relevant. It is up to the judging subjects themselves to decide, under improved epistemic conditions, which things they *deem* to be morally relevant.

Remarkably, some philosophers who favor a strong version of emotionism subscribe to a set of qualifications similar to the one just presented. Jesse Prinz, for instance, cites the evidence from psychopathy and acquired sociopathy as well as filthy-desk-style examples in favor of both the necessity and the sufficiency thesis. But he implicitly rejects it at the same time. Following his discussion of the empirical findings about moral judgment, he writes:

> The first problem [of one form of sentimentalism] has to do with error. If "wrong" referred to whatever causes disapprobation in me, then I could not judge something to be wrong in error. To avoid this consequence, we must idealize. We should say that the word "wrong" refers only to those things that irk me under conditions of full factual knowledge and reflection, and freedom from emotional biases that I myself would deem as unrelated to the matter at hand. (Prinz 2006, 35)

Clearly, any reasonable agent has to deem picking up on an arbitrary trigger word as unrelated to a moral issue. The trigger word is *arbitrary*, after all.

Genuine moral judgments "are those that are regulated or endorsed by reflection" (Kennett and Fine 2009, 78). What the empirical evidence shows, however, is that cases where a moral judgment is *regulated* by reflection from the outset are rare, and throughout this book, I have argued that these cases are the exception. Reflective regulation only occurs, and is only required to occur, when a moral intuition has been legitimately challenged. In this case, an appropriate response is required, and coming up with such a response will take some reflective effort. Our educated moral intuitions are not regulated by conscious online reflection; they arise quickly and effortlessly, because responding with this type of intuition to this type of situation has become automatic over time. But although the vast majority of our moral judgments are not regulated by reflection in this direct sense, there still is an important connection to episodes of reflection. For one thing, our moral intuitions are acquired over the course of our moral education in which explicit moral reasoning plays an important formative role; for another thing, our moral intuitions remain susceptible to explicit moral reasoning even after they have become automatic.

It thus seems more promising to work with *counterfactual* reflective endorsement. Competent moral judges need not arrive at their judgments by a causally effective conscious process of reflection, but are allowed to arrive at them via an emotionally triggered intuitive process that is reflectively acceptable. By and large, we can say that proper moral subjects reconsider their initial judgments after being informed about their causal genesis. If a judgment survives this scrutiny, one can say that the subject *reflectively endorses* it.

Note that there are two possible objects of endorsement here: the moral judgment itself, and the "method" by which the subject arrived at it. My main focus lies on the latter, because I am talking not about the conditions under which moral judgments are true (or otherwise correct), but about the conditions under which emotional reactions count as *genuine* moral judgments. If, by accident, a subject arrives at a correct judgment using a fluky method, she can still endorse the judgment and stick to it. But she will have to deem the way she arrived at it inappropriate because it was fluky.

Here, then, is why I think the emotionist challenge to rationalism about moral judgment can be dissolved: emotions are *necessary* for moral judgment because moral judgments without feelings are empty, and feelings without moral reasoning are blind. We are now in a position to see that the fact that emotions are *sufficient* for moral judgments need not threaten the rationalist, either. Emotions only produce *genuine* moral judgments, rather than mere reactions of disgust or compassion, when they pick up on the morally relevant features of the situation. But when they do, they also confer sensitivity to moral reasons on the judgments they yield. Rationalism requires nothing more.

This is the main line of my argument. It is difficult to provide an answer to the question under which circumstances a subject is rationally entitled to endorse the "method" by which she arrived at it as relevant, but I wish to make an attempt at developing such an answer. What I want to argue for is the following simple but powerful idea: a subject can reflectively endorse her judgment if the method she used to arrive at it is sensitive to the morally relevant features of the situation. This sensitivity, or reasons responsiveness, is specified in counterfactual terms. A method is sensitive with respect to X if and only if, under slightly different circumstances (or, perhaps more technically, in close possible worlds) where X obtains, the method would lead a subject to believe X. Conversely, under slightly different circumstances where X doesn't obtain, it would lead the subject to acknowledge this as well.

Reasons responsiveness is a normative concept. It presupposes certain standards of correctness. We are interested in sensitivity because reasons-sensitive methods increase the likelihood of subjects arriving at *correct* results. What else could it be that justifies my accepting the way my moral belief was brought about? Since as a sincerely morally judging subject, I am interested in the correctness of my judgments, I can endorse the way

I arrive at them if the method used primarily serves that particular purpose.[3]

Two important questions have to be addressed. First: what is it that makes a method reasons-sensitive? And second: does the above discussion show that while the empirical evidence supports simple emotionism′ it is ruled out by sophisticated emotionism? Does it show that the way subjects arrive at their moral judgments in filthy-desk-style scenarios is irresponsive to morally relevant and responsive to morally irrelevant factors?

As for the first question, I would like to return to an example from above. If I witness an old woman being mugged by a young man, I am observing a chain of events that automatically elicits a reaction of empathic distress and presumably anger. Provided that I do not come across additional information about the incident that mitigates my reaction (if it turns out, for instance, that the old woman is actually a young man, disguised as an old woman, and that he stole the other young man's purse right before), I will take my emotional reaction—as the evidence for the necessity thesis predicts—as a challengeable "ought not-appearance" and judge that what the young man did was wrong. Can I reflectively endorse the method with which I arrived at my verdict? According to the above proposal, I am entitled to do so if it is responsive to something of moral relevance, and thus likely to produce correct moral judgments even under slightly different circumstances. Under normal circumstances, observation of a *prima facie* harmful incident is likely to cause emphatic distress. This feeling of distress will "pick up" on the morally relevant features of the situation, and confirm that a harmful action has in fact been done. It responds to the fact that a presumably innocent person is being stolen from without good reason, and the disposition to respond to such acts in that way reliably, though not infallibly, leads to a correct normative attitude.

In most cases such as these, we can work with an intuitive notion of moral relevance that remains noncommittal with respect to competing normative as well as metaethical theories. Regardless of one's favorite account of normative ethics, one can agree that considerations of the harmfulness are pertinent to its rightness or wrongness, whereas considerations of how unpleasantly the environment smelled in which the respective judgments were made are not. One need not be a realist about value to defend such a notion of moral relevance. Even error theorists can agree that we can offer criteria to distinguish subjects' moral from their nonmoral judgments. There are some things which people *take to be* morally relevant. For this to be the case, there need not be anything out there "in the world" to ground this distinction.

As for the second question, the task is to see whether *Sophisticated Emotionism* can successfully cope with cases that, intuitively, don't seem to meet the constraints on proper moral judgment. Subjects in the disgust condition of an experiment (for instance, people who are placed behind a filthy desk) predictably judge morally wrong actions more severely, and people under the influence of hypnosis may judge actions wrong that aren't morally wrong at all. But what these people respond to in experiencing an emotional reaction—such as the word "often" in a story vignette—is not a morally significant feature of the situation at hand. Imagine a person that judges a morally wrong act to be wrong, but does so by picking up on a trigger word used in the description of the story. Clearly, that person would have judged a different action to be wrong as well, even though it might not have been, as long as it had contained the trigger word. The method that person has used to arrive at her judgment does not reliably "track" moral wrongness. A person that reacts outraged upon reading an article about female mutilation does (provided this reader is not under the influence of hypnosis and the like).

Take the following case, devised by Gilbert Harman (2007): you witness a group of children setting a cat on fire for fun. They pour gasoline on it, and ignite it. Now suppose you are participating in a psychological experiment and a story vignette is presented to you that contains a description of the incident. It might read something like this:

Scenario I

You walk around the corner and see a group of young people. You witness one of them catch a cat, another one hold it to the ground, another one pour gasoline all over it and another one pull out a matchbox and set it on fire. They observe the cat while it tries to escape its misery; occasionally, they laugh at the animal's screams and its futile attempts to survive, and they stay until the charred, dead body stops burning.

Most people will be horrified by imagining this scene, chilled by the youngsters' callousness and outraged upon their wanton cruelty. The judgment that what these young people did was wrong literally forces itself upon us. Now suppose that, via posthypnotic suggestion, you have been primed to feel a pleasant warm feeling upon reading the arbitrary trigger word "often." You are now presented with a slightly different version of the story (call this version *Scenario II*) where the trigger word has been inserted into the last sentence. Due to the fact that your feelings have been manipulated through hypnosis, the judgment that what they did wasn't all that bad replaces your response. You don't know why, but what these youngsters did suddenly seems quite kind to you. Unbeknownst to you, you have not responded to the morally relevant features of the situation, but to a trigger word that you have been primed to pick up on. But clearly, using that method proved to be unreliable. Under slightly different circumstances—only the word "often" has been added to the story, after all—where the judged action is still wrong, or at least no different in all important respects, you diverted from your initial judgment and changed your mind. But the described action *is* wrong. It is the unreliable method you used in arriving at your judgment that made you think otherwise. Moreover, a competent moral judge would thus dismiss the method upon reflection under ideal conditions in which he has been informed about the causal genesis of his judgment.

Whether a subject has arrived at her judgment using a method she can reflectively endorse and whether (although the subject didn't use a reasons-responsive method) she is prepared to reconsider her judgment depending on how it fares under reflective scrutiny, are two different things. Given its openness to rational reflection, however, the second case ought to be classified as proper moral judgment as well. This, I suggest, can be disjunctively accounted for along the following lines, call this *sophisticated emotionism′*:

Sophisticated emotionism′

It is necessary and sufficient for making a moral judgment to be disposed to have an emotional attitude of moral (dis)approbation towards an action that either

(i) causes the judgment in a way that *could* be reflectively endorsed by the judging person under ideal conditions of full information and rationality, or
(ii) would allow the subject to withdraw her judgment in the light of undermining evidence, or
(iii) to back it up with appropriate further grounding.

In a nutshell, my response to the emotionist challenge is that sophisticated emotionism′ is compatible with, and indeed a form of, moderate rationalism about moral judgment. But it is a rationalism with a human face that does not reject the impact of emotions as epistemically problematic and impure.

It makes sense to classify judgments that fall under the second and third disjuncts as genuine moral judgments, too, because although in (ii) and (iii), a subject does not at first respond to a morally significant feature of the situation she judges, she nevertheless possesses the cognitive virtues that are needed for genuine moral competence that, in a way, is another reliable method of forming moral beliefs. We are all prone to be misled, tricked, and fooled by our emotions. Whether one engages in the practice of genuine moral judgment depends on how one deals with this fact.[4]

One can also make the above point the other way round. Suppose the test subjects are psychopaths—who typically score low on susceptibility to empathic distress. You confront them with *Scenario I*. They react pretty untouched and judge the case accordingly. As we have seen, they lack the necessary "perceptual" capacities to pick up on the morally relevant features of the issue at hand. Psychopaths are, however, not deprived of all affective dispositions. So you prime a control group of psychopaths to respond with disgust to the trigger word in *Scenario II*, and the subjects in that group judge the children's violent acts to be wrong. Even though in the latter case, the emotional response did in fact lead to a correct moral judgment, the method that was used cannot be reflectively endorsed under conditions of full information and rationality. Responses to extraneous features of a situation do not guarantee that one's judgments track the available moral reasons.

On the other hand, a morally competent judge will respond to relevant features of the situation in question, and his judgments will not be subject to manipulation by random influences. Suppose a normal, healthy, unhypnotized adult individual is presented with a third story that frames roughly the same chain of events as in *Scenario I* in a completely different way. We can expect a normal person—a person who thinks what these kids did was terribly wrong—to be insensitive to the changes in wording between the first and the third scenario when making her judgment, because these are features she did not—and should not—respond to in making her judgment. Rather, her judgment was caused by her empathic distress, her feelings for the suffering animal. This is how she arrived at her judgment, and due to the fact that this method is likely to lead to correct judgments, it conveys the justificatory force necessary to render a judgment a moral judgment, and a moral judge a competent one.

Sensitivity to morally relevant considerations is about which emotional processes that cause the subject to make a certain moral judgment convey justification on that judgment. The distinction between processes that have this kind of normative force and those that do not is needed to capture the intuition that reactions of, say, empathy or disgust which *merely happen to be* about morally salient scenarios, but do not pick up on the morally relevant features of those scenarios, do not count as genuine moral judgments. Reasons sensitivity is about the features that ground this reflective endorsement. Suppose a subject has formed a moral judgment about an action X, but has done so using a "method" that—oblivious to her—picks up on an arbitrary feature of the situation, such as ambient noise, a bad smell, or a trigger word. This method is unreliable not because it is insensitive to the moral features there are (because it cannot be said to "track" them), but in a way that is internal to the subject: using the same method, the subject could and would have formed the opposite judgment about the case at hand if her emotional state had been manipulated differently; and she would have made the very same judgment, given the same kind of manipulation, even if the case had been entirely different. That fact alone rules out the method as untrustworthy, because it is not sufficiently robust and insensitive to arbitrary features. This lack of robustness can be detected simply by comparing two different scenarios. One need not invoke an objective standard of moral truth here.

If the subject came to know about how she arrived at her judgment, would she still accept it? In arriving at her judgment, has the subject been causally influenced by features she deems relevant to the issue in question? The account is not committed to a strong form of moral objectivism: there are no moral facts in a straightforward sense, over and above the emotional reactions we would have under improved conditions. The account also holds that it is in order to reflectively endorse an emotional process by which one has arrived at a moral verdict if and only if that process responded to, or is responsive to, the morally relevant features of the situation, action, or person the judgment is about.

Picking up on morally irrelevant features cannot be endorsed because doing so can—as the aforementioned studies demonstrate—lead one to form a judgment one might not be willing to accept in the light of information about its actual causal genesis, unless one can back it up with appropriate further grounding that *reconnects* one's judgment with features one *could have* responded to in arriving at one's moral belief. To say that there are morally relevant features is not to say that there are moral facts. Moral facts are mind independent facts about what is morally right or wrong. I do not hold there are such facts. What is morally right or wrong is determined not by facts but by the rationally amenable emotionally charged educated intuitions we have. The considerations that are relevant, in a domain-specific way, for the normative assessment of those reactions consist of ordinary facts together with considerations that refer to the morally relevant features of a situation. That what is morally relevant or not—considerations of harm or fairness, for example—is also due to the emotional capacities of human beings does not render the proposal incoherent. It is a trivial fact without which morality would be pointless to begin with.

What those features are, and what renders a feature morally relevant or irrelevant, is a question for normative inquiry. A possible suggestion, however, would be to start from intuitively compelling examples and work one's way up from those distinctions. (Bad smells and filthy desks are paradigm examples for features that are morally irrelevant; violations of accepted moral principles and considerations of harm are good examples of things that are morally relevant.) What counts as morally relevant and irrelevant is still under negotiation: it is both essentially contested and historically variable. Disgust, for example, can pick up on arbitrary trigger words, the pollution of a sanctuary, or the description of an appalling violent crime. Whether only the third or also the second thing are morally relevant is a normative question, a question about what one ought to do. But the distinction between morally relevant and irrelevant factors alone involves no commitment to metaethical objectivism. The story told in this chapter is thus not only a story about what moral judgment is, and what the psychological basis of genuine moral judgments is, but also a story about how metaethical questions naturally lead into questions of normative ethics.

4 Conceptual Constraints

Are emotions necessary and sufficient for moral judgment? I have argued that they are, but that a moderate rationalist about moral judgment does not have to worry about this. Emotions are necessary for moral judgment in the same way as perceptions are necessary for judgments about the external world, and emotions are sufficient for moral judgments only if they cause them in a normatively acceptable way.

One might worry that the argument I have presented uses a conceptual magic trick, designed to simply dismiss the empirical evidence for the importance of emotions for moral cognition by "fiat." It seems that the conceptual argument just presented offers a refutation of empirical models of moral judgment simply by saying that what is studied in the respective experiments isn't genuine moral judgment *by definition*. This is not what I want so say, and it wouldn't be a plausible objection anyhow. When people are judging cannibalism or incest, they really are making (deficient) moral judgments, even in cases where it can be shown that their judgments are nothing more than expressions of disgust. Compare the aesthetic judgment by a person who thinks Otto Dix's triptych *The War* is "ugly" because it depicts "ugly" things—shredded bodies, burning land, soldiers with gas masks. This may well be poor reasoning, but still: we have to admit that it is an aesthetic judgment, although a deficient one. I suggest describing the moral case in the same way: people whose judgments are not responsive to relevant moral considerations, but are triggered by uncontrollable and hence unreliable emotional responses, are making deficient moral judgments. They may be *trying* to make moral judgments, they *think* that they do, but if they show themselves not to stand in any kind of autonomous—that is, rational—relation to their judgments, they cannot be said to engage in the practice of moral judgment *in the right way*. Constantly failing to live up to these standards renders a subject morally incompetent, and ultimately deprives it of its status as a morally judging subject altogether (although it remains the legitimate addressee of obligations and the possessor of rights, of course). A full-blown moral judge must eventually take an interest in the trackingness (as specified in the first disjunct of the above account) and rational answerability (as specified in the second disjunct) of her moral beliefs.

It remains an open question whether the account developed in this chapter describes what subjects actually do or whether it specifies what agents ought to do. This is not an accident: the argument above is supposed to be psychologically realistic and empirically adequate; at the same time, however, it is also supposed to be demanding enough to make normative claims on moral agents, claims that these agents can fail to meet. It can be shown empirically that subjects do engage in normative reflection about their immediate emotional attitudes and automatic intuitions (Schwartz and Clore 1983), and that they are prepared to discount distorting influences on their judgments and correct for extraneous influences to their beliefs, once these are made accessible to them. The influence of filthy desks on people's moral judgments can be seen as such a form of unwanted mental contamination (Wilson and Brekke 1994).

In fact, the above conditions specify norms of moral competence test subjects themselves hold to be adequate. The findings by Wheatley and Haidt about the connection between hypnotic disgust and the severity of moral judgment nicely illustrate that certain standards of correctness are not externally imposed on people by philosophical theory, but are actually written into the patterns of moral reasoning by ordinary subjects themselves. Remember that the research conducted by Wheatley and Haidt used the method of posthypnotic suggestion to prompt people to experience a quick flash of disgust after hearing (or reading, respectively) an arbitrary word such as "often." Here is how some of the subjects described their experience:

> "When 'often' appeared I felt confused in my head, yet there was turmoil in my stomach. It was as if something was telling me that there was a problem with the story yet I didn't know why." One nonamnesic participant commented: "I knew about 'the word' but it still disgusted me anyway and affected my ratings. I would wonder why and then make up a reason to be disgusted." (Wheatley and Haidt 2005, 783)

People are confused about an emotional reaction for which they can find no plausible source. They hesitate to make the judgment they are inclined to make, and rightly so. They even admit that they would "make up" a reason, implying that they know that there actually is none. Subjects are aware of the fact that their response to an irrelevant cue does not reliably indicate the moral wrongness of an action or the blameworthiness of a person.

5 What Reflection Can (and Cannot) Do

Emotions are sufficient for moral judgment only if this causal sufficiency is justification-conveying. A moral judgment which was caused by an emotional reaction toward ambient noise or an arbitrary trigger word does not meet this standard, as these extraneous influences could not be endorsed upon reflection: hypnotically induced disgust is an emotion that, as far as its causal genesis is concerned, did not arise in response to features of the judged situation that are morally relevant.

This interpretation of the sufficiency thesis makes use of the notion of reflective endorsement to distinguish genuine moral judgment from mere reactions of disgust or attraction. Reflective endorsement is one possible way for human moral judges to increase their reason responsiveness. As the evidence discussed above shows, subjects can be made to pass moral verdicts on the basis of attitudes whose emergence bypasses their sensitivity to morally relevant factors, and exploits subjects' proneness to be influenced by extraneous, but emotionally gripping situational aspects. Moral reflection can be a remedy to this proneness.

In what follows, I wish to consider two arguments that attempt to show that the concept of reflective endorsement in general is not as helpful as it may seem. The first argument charges analyses of concepts that operate with idealized conditions with committing a conditional fallacy (Johnson 1999; van Roojen 2000; Gert 2002). This argument claims that reflective endorsement is of no help whatsoever. The second argument, which has been championed by Hilary Kornblith (2010), attempts to point out that the usefulness of the concept of reflective endorsement under ideal conditions is modest at best, and that the judgments subjects come to endorse upon reflection are not necessarily better than those they do not. This argument claims that reflective endorsement is only of very little help. Let me take up these arguments in turn, and show why the above account remains untouched by them.

My argument understands genuine moral judgments in terms of how one would behave if one were in ideal conditions. Responding to disgusting smells does not constitute a genuine moral judgment, because under ideal conditions, one would not respond to disgusting smells. And, once a subject has become aware of this extraneous influence under ideal conditions, the subject would give up the judgment or back it up with further grounding. But one problem that looms for this account is that the *analysandum*—what a genuine moral judgment is—might not be correctly explained by the conditional in the *analysans* (if one were in ideal conditions, this is what one would do), because typically, the conditions which are specified in the antecedent of the *analysans* make behaving in such a way as the consequent specifies pointless—if I were fully rational, I would not back up or reconsider my judgment, because that would not be necessary (because I would be fully rational). It might be rational for me to seek psychiatric help. But if I were fully rational, I would not need this help. Therefore, the analysis tells me that I do not have a reason to seek psychiatric help. But I do. This is called the "conditional fallacy" (Shope 1978).

Conditional analyses like these have become especially popular in theories of practical reason. Many authors, for example, try to cash out what it means to have a reason to do something in terms of what an ideally rational and fully informed person would want to do. I do not have a reason to drink a glass containing a liquid that I think is water but really is petrol, because a fully informed person would not want to drink the liquid. The problem is that there are cases in which, intuitively, one does have a reason to do something, yet the rational person would not want to do that thing—so the analysans cannot be correct. Michael Smith (1995) discusses an example in which a man has lost a squash game, and contemplates leaving the court immediately because he is afraid he will lose his temper, become angry, and hit his opponent. The conditional analysis falsely says that this man does not have a reason to leave the court, because a fully rational person, being in full control of his temper, would want to do no such thing.

To find out whether this problem applies to a counterfactual analysis of moral judgments as well, one would have to come up with a case in which, intuitively, one is clearly making a moral judgment (although maybe not a correct one), yet that judgment is one that a fully rational person would not make. Say a subject *S* comes to believe he should leave his family because under hypnosis, it has been suggested to him that this would be a good idea. A fully rational person (call him, using standard notation, *S+*) would not endorse this method of arriving at a moral decision as relevant to the question at hand. Therefore, it does not constitute a genuine moral judgment. The analysis works for cases like this. But now consider a second example: Say *S* comes to believe that he should leave his family because he is afraid he will be unable to control his pedophilic urges and molest his newborn daughter. This seems to be a clear case in which someone makes a moral judgment on the basis of considerations which are indeed genuinely relevant. Yet this is a method of arriving at a moral verdict that a rational person would never use, either because that person would not have such urges in the first place, or because the person would be strong-willed enough to be able to guarantee not to act on them.

There are two possible responses to this problem. The first one draws on Michael Smith's (1995) advice model of practical reasons. This model has it that if one has a reason to do *X*, then *X* would not be the action that *S+* would (want to) perform herself, but which *S+* would *advise S* to perform. On this advice model, one would be making a genuine moral judgment if one could reflectively endorse the way in which one arrived at the judgment under ideal conditions; to endorse a method in such a way, however, does not mean that a fully rational person would use this method herself, but that she would advise her less than ideally rational counterpart to use the method. This proposal circumvents the conditional fallacy.

To be sure, the advice model of reflective endorsement has problems of its own, but they only occur in the context where the *analysandum* is "having a reason." What one has reason to do has an explanatory component (Johnson 1999): if something is a reason for someone to do something, then that reason must, in principle, be capable of swaying that agent, of explaining his actions. But the advice model violates this requirement. If a reason is that thing which an ideal agent would want to do, and if the thing that the ideal subject would want is that the nonideal version should have a desire, then the latter—the desire that someone else should have some desire—cannot explain the less ideal agent's actions—it's just not that kind of thing. The simple ideally informed desire to do X could play this explanatory role, but then again, this is the very *analysans* that ended up in a conditional fallacy. But the good news is that none of this applies to the advice model of reflective endorsement *in the case of moral judgment*. Because here, we are talking about what it means to make a moral judgment, not what it means to have a reason. And the concept of a moral judgment is not subject to an explanatory requirement of the kind just mentioned, because an account of what moral judgments are does not have to causally explain why and how agents come to be motivated to act.

The second way one might overcome the obstacle posed by the conditional fallacy is to bear in mind that it is not the case that an emotional reaction to a morally salient scenario only counts as a genuine moral judgment if the judgment were one that a fully rational person would make. This could work as an account of what a *correct* moral judgment is, and this is not what I am interested in here. I am interested in what makes a *mere candidate* for a moral judgment a *genuine* moral judgment, and I am primarily focused on finding an *analysans* that explains why picking up on arbitrary trigger words and filthy desks does not count as an exercise of moral competence. Clearly, a fully rational person would not make the moral judgment to leave his family because of his uncontrollable urges, because a fully rational person does not have such urges (or could easily overcome them). As an analysis of what a correct moral judgment is, that analysis would therefore commit the fallacy, because given his circumstances, the nonideal, pedophilic subject does, it seems, make a *correct* moral judgment. But S+ could, of course, *endorse the method* by which S arrives at his judgment, because making a decision on the basis of a caring emotion toward one's loved ones is a response to a morally relevant factor—the potential harm inflicted upon the child and the whole family. Either way, the account turns out adequate. Reflection upon improved conditions should be seen as a standard against which to evaluate one's judgmental behavior. Sound methods of judgment formation can be counterfactually endorsed, even when the judgments themselves cannot. When no sound method has been used, and this is pointed out to a subject under *actual* improved conditions, a morally competent subject needs to react accordingly, and suspend judgment or adduce further reasons.

The second worry about reflective endorsement accounts I wish to defuse comes from Hilary Kornblith (2010), who has recently argued that reflection cannot do the job it has been hired to do. This is not, as the arguments surrounding the conditional fallacy have it, due to the fact that reflective endorsement analyses yield incorrect results, but because the importance of reflection is grossly overestimated. Or so Kornblith argues, anyway.

Kornblith starts from Sosa's (1991) idea that there are two types of knowledge, animal and reflective knowledge, and that the latter type is "better," that is, more justified. Sosa defines the methods of belief formation operative in cases of reflective knowledge as processes that, were they to be exposed to additional evidence to the contrary, would lead the subject to change her mind. But Kornblith rightly points out that, although sensitivity to evidence is indeed an intellectual virtue, it is what animal minds have as well, and so one needs to operate with a stronger notion of reflection, a notion according to which a reflecting subject actually needs to stop engaging in first-order methods of ordinary belief acquisition, and explicitly reflect, from a second-order perspective, on the first-order beliefs she has and the means by which she has arrived at them. However, whether this type of epistemic behavior is recommendable is a genuinely open question: "If what one cares about is the reliability of one's process of belief acquisition, then reflecting on one's beliefs is a mixed bag: sometimes better, sometimes worse and sometimes just the same as belief uninfluenced by subsequent reflection" (Kornblith 2010, 4). As we have seen in the introduction, it is an empirical fact that conscious reflection and explicit reasoning do not always improve one's cognitive performance. The above account should be able to incorporate this fact.

Some authors grant an even stronger role to reflection, most prominently Christine Korsgaard (1996, 2009), who argues that despite the empirical limitations of episodes of reflection, reason in general and reasons in particular are grounded in reflective endorsement. This, Korsgaard argues, is due to the reflexive structure of the human mind, for which nothing that has not been approved by reflection can constitute a full-blown reason. Kornblith disagrees with this view, and here's why. He asks you to imagine that you have a belief. Then, for some reason, you ask yourself "the normative question" (Korsgaard 1996) whether you have reason to hold it. You start reflecting and bring various rational standards of belief-evaluation to bear upon the belief (say that your friends have told you your belief is true, it very strongly seems to you to be true, you notice that you have been right about matters like these before, you read about it in the newspaper, you can see that it is the case, you remember that it always has been the case, and so forth). After having so reflected, you end up endorsing the belief. Do you now have reason to stick to the belief? Not obviously, because although in this case everything went fine, sometimes we apply *bad* standards of belief evaluation. Reflecting on your beliefs cannot guarantee that you don't, and therefore it does not constitute reasons for belief (or action, for that matter), which is precisely what it was supposed to do. We ought to believe *p* only if *p* is true, not if it passes reflective scrutiny.

This is of course correct, but the point about reflective endorsement is a different one. It is that when you really are in doubt, what can you do? Simply saying that you ought to believe that which is true is correct, of course, but it doesn't help very much, does it? Because this is the problem—you have lost your confidence in what is true, which is why you have started to engage in reflection in the first place. Now reflection, and the endorsement you arrive at after you are done reflecting, does not guarantee that you will end up with the correct result. Reflection is a means of belief acquisition just like any other, and so just as fallible. But all of this merely entails that finding truth and justification is very difficult, not that reflection and reflective endorsement are utterly useless. Kornblith's argument is fine, but he overestimates the implications of its conclusion because he overestimates the significance of his target: most people are not committed to saying that reflective endorsement is a *source* of normativity, let alone the only one. But it is one more way for us to become reason responsive. And in many cases, when people have to take into consideration new information, especially when it is relevant to whether they should correct their judgments or leave them uncorrected, there is, as an empirical matter of fact, no alternative to it. All too often, we are mere leaves in the wind of our unconscious; if reflection helps improve our situation in this regard, its assistance should be welcomed (Tiberius 2013).

6 A (Limited) Role for Disgust?

The influence of disgust on moral judgment makes many people—myself included—at least somewhat uncomfortable. Some reject it outright, arguing that there is little reason to assume that what is colloquially labeled *moral* disgust should count as genuine disgust (Gert 2015). As far as the empirically demonstrable influence of nonmoral disgust on judgments of rightness and wrongness is concerned, however, such doubts strike me as beside the point.

More on target are others who argue that disgust can be morally relevant, because it tracks social contamination in just the way ordinary nonmoral disgust tracks the contaminating presence of germs and parasites (Plakias 2013) or because disgust responses tacitly embody hidden principles that can facilitate moral knowledge (Wielenberg 2014). Even for those with decidedly liberal moral views, there seems to be little reason to grant disgust no role whatsoever in the production of moral judgment (Kahan 1999).

One important argument against the moral relevance of disgust turns on how the mechanism of disgust has been designed to work (Kelly 2011). In order to properly perform its function of pathogen avoidance (Tybur et al. 2013; Strohminger 2014), disgust needs to be a "better safe than sorry" mechanism; by and large, the comparatively low costs of missing out on an opportunity for food are preferable to the potentially existential costs of ingesting something toxic. This leads to an unhealthily high *false positive* rate: many things that aren't actually contaminating get classified as such; in the moral domain, this often has disastrous consequences, for instance when actions or groups that frequently evoke disgust—such as unusual sexual practices or unfamiliar looking people—are condemned and ostracized.

Note, however, that although this argument may show that disgust is morally unreliable because it is too trigger happy, it does not entail that this unreliability stems from its moral extraneousness. Disgust is morally *hypersensitive* rather than irrelevant. By analogy, it seems implausible to assume that empathy—which, as recent evidence suggests, is subject to troubling phenomena such as compassion fade (Västfjäll et al. 2014) and the perception of pseudoinefficacy (Västfjäll, Slovic, and Mayorga 2015)—is morally irrelevant because it is *hypo*sensitive (Singer 2005; cf. Prinz 2011b). At most, Kelly's argument shows that we should be especially wary of the moral impact of disgust, but not that we should rule it out it completely. Moreover, disgust plays an important explanatory role in underwriting one of the six moral foundations the asymmetrical endorsement of which accounts for differences in political views between liberals and conservatives (Sauer 2015).

From the perspective of the educated intuitions model, one problematic feature of disgust is its apparent lack of plasticity. But this feature, too, is typically exaggerated, as diachronic as well as cross-cultural evidence suggests that in addition to things such as feces or rotten meat that seem universally disgusting, disgust is no less malleable than other emotions. Many things used to disgust us but no longer do, and many things used to be considered normal but now seem gross (for one striking example, see Rozin and Singh 1999).

When, and only when, disgust picks up on the morally relevant features of the situation and doesn't pick up on the morally irrelevant features of the situation, it can be used as an (admittedly fickle) guide to moral knowledge. But how strong is the influence of disgust on our moral cognition, really? According to the standard narrative often propagated, feelings of revulsion have a deep and lasting role to play in morality. But a close look at the empirical findings reveals that support for this claim is surprisingly weak.

First of all, it should be mentioned that several of the studies on the effect of artificially induced disgust or mood changes on moral judgments mentioned earlier in this chapter did not always replicate (Johnson, Cheung, and Donnellan 2014; Earp et al. 2014; Duke and Bègue 2015). Then again, this does not necessarily mean that these studies are flawed. After all, a replication effort is just another study that may not replicate itself. Together with the original study, it becomes part of the overall case for and against the existence of an effect. Perhaps more importantly, a recent meta-analysis of (published and unpublished) studies of the amplifying effect of disgust on people's moral beliefs found the overall effect to be fairly small ($d = 0.11$; Landy and Goodwin 2015). Moreover, there may be a file drawer problem at work here, as the effect disappeared completely when only unpublished studies were analyzed, suggesting that even the small overall effect size found may be the result of publication bias. Finally, even if the findings turned out to be robust in the end, it would be a gross overstatement to say that they show moral judgment to be based on disgust. Most of the studies on incidental disgust generalize from rather specific subgroups (such as highly hypnotizable or easily disgusted subjects); effects are found only for a subset of used vignettes; and, as I have mentioned earlier, none of the studies change the polarity of people's moral judgment from *right* to *wrong* (May 2014). The effect is an amplifying one that affects only the *severity* of people's moral judgments. And this is still a very far cry from showing that feelings of disgust are sufficient for moral cognition.

Conclusion

This chapter concludes my response to the emotionist challenge. This response is animated by the idea, already set out in earlier chapters, that in human moral judgment, emotion, intuition, and reason go together. This is true in a twofold sense: first, because emotions form part of subjects' practical reason; and second, because emotion is aided by reasoning—and vice versa—in the production of moral judgments.

The particular results obtained in this chapter consist in a reinterpretation of the empirical evidence for the claim that emotions are sufficient for moral judgment. This sufficiency thesis seemed to threaten the rationalist project, because the evidence for it suggested that the emotional reactions people's moral judgments are expressive of are very capricious indeed. And while it might be that reason and emotion are not *in principle* incompatible, the *de facto* basis of moral judgment on something so volatile did not seem to be of much help to the rationalist.

My response to this problem was to distinguish between two different kinds of sufficiency. I argued that the experimental evidence manages to show that emotions are *causally* sufficient for moral judgments—changes in subjects' emotions often account for changes in their moral beliefs. However, unless the emotional processes that cause the judgments do not retain at least some sensitivity to the morally relevant features of the situation, they do not yield genuine *moral* judgments anymore, but collapse into mere reactions of disgust, anger, or contagious distress, which even anti-rationalist emotionists want to distinguish conceptually from proper moral attitudes. And when they retain this sensitivity, my argument proceeded, the aforementioned worry about the dependability of moral judgment disappears.

I have suggested one possible way one might try to cash out the distinction between morally relevant and irrelevant factors, and I have used the notion of reflective reasons responsiveness to do so, for various reasons. One is that it fits into a successful tradition in metaethics and moral philosophy, a tradition that members of both the Humean and the Kantian tribe can feel comfortable with. Another reason is that the reflective endorsement account nicely coheres with the *challenge and response* model of moral reasoning sketched in the first part of this book. Moral agency, that is, the capacity to make and act upon moral judgments, requires one to respond intelligently to relevant information and to remain open to hitherto unacknowledged considerations. Moral reasoning is how subjects exercise this capacity, and reflective endorsement or rejection of one's moral beliefs are the two main ways to resolve the tension that comes with new challenges.

There might be other terms in which to understand morally relevant considerations. My point is that there is no morally neutral way to identify the subject matter of moral psychology without at least some preconception of what makes a judgment about an action, person, or situation a distinctively *moral* one, and any such preconception will be, to some extent at least, normatively laden. Consider the psychology of perception as a comparison case: one cannot even pick out perception as an object of study without assuming that perception is the thing that gives perceivers sensual information about the external world, and this preconception is not entirely neutral with respect to the epistemic status perceptual episodes enjoy within the cognitive economy of living organisms. Similarly, one cannot pick out the subject matter of moral psychology without making some assumptions about what does and does not count as morally relevant. It is important to emphasize that a commitment to moral realism or objectivism is not among those assumptions. One can rely on what even the staunchest antiobjectivists about morality themselves consider to be morally relevant: some think that considerations of purity are, some think community and authority are moral concepts. Nobody, however, thinks that sounds, smells, and arbitrary words are among the morally relevant features of the situation. It is this distinction my account is trying to capture. The distinction is real; whether the account I propose adequately captures it is, of course, a different question.

III Applying the Model

Consider the following inference: no addictive things are cheap; some cigarettes are cheap; therefore, some addictive things are not cigarettes. If you buy into this conclusion on the basis of the premises, rather than on independent grounds, then your intuition has fooled you—as it does most people. Unless we have received special training, we routinely—and mistakenly—assess the validity of arguments on the basis of prior assessments of the believability of the conclusion (Evans 2008).

Why does this happen? Like moral dumbfounding, this "belief bias effect" and many other errors in thinking result from the fact that cognitive processes are carried out by two mental subsystems: System I and System II. Your intuitive System I tells you that the conclusion is believable; this leads you to the judgment that the argument is probably valid. But you are mistaken, as the truth of a conclusion and what it follows from are two entirely different things. If you reflect on it for a little longer, your reflective System II might tell you so as well, and will make it possible for you to correct your initial mistake.

In this chapter, I bring the account I have developed in this book to bear on what has come to be known as the *dual process theory of moral judgment*. This theory studies the way the intuitive System I and the reflective System II interact not in how people reason deductively, but in how they arrive at moral judgments. Its main claim is that our intuitive system distorts our moral judgments in roughly the same way as it biases our reasoning: by making us jump to beliefs which seem superficially plausible, but are actually unjustified. I will revisit the link between emotion and moral judgment, the automatic basis of a large portion of our moral cognition, and the possibility that at least some of our efforts at conscious moral reasoning consist in mere *confabulation* rather than genuine *justification*. This third part ties my responses to the anti-rationalist and the emotionist challenge to moral rationalism together in one single thread. It can be seen as a case study in why automaticity and rationality in moral judgment are not incompatible with each other and, at the same time, a useful test case for the viability of the educated intuitions model.

Here is how I will proceed. Psychologists and cognitive neuroscientists such as Joshua Greene (Greene et al. 2001, 2004) have argued that there is empirical evidence that automatic processes are essential not for moral judgment in general, but for one particularly important subclass of moral judgments: so-called *deontological judgments*. Traditionally, these have been accounted for by Kant and Kantian moral theory. This thesis runs counter to most of the philosophical mainstream, which takes consequentialism to be the natural ally of metaethical sentimentalism (because emotional concern for the well-being of others seems to be what makes us susceptible to the normative significance of utilitarian considerations), and positions that stand in the Kantian tradition to cohere best with metaethical cognitivism (because acknowledgment of rational principles seems to be what makes us susceptible to the demands of pure practical reason).

That, at least, is what Kantians would like us to believe. According to Greene, consequentialist moral judgments—roughly speaking, judgments whose primary concern is the maximization of utility in a given case[1]—are based on controlled rather than automatic cognition, and in making them, we recruit rational capacities rather than intuitive feelings. On the other hand, deontological moral judgments—roughly speaking, judgments whose primary concern isn't the maximization of utility, but the moral assessment of actions in terms of their intrinsic moral worth—are based on automatic responses rather than cognition, and in making them, we recruit emotionally valenced intuitions and feelings rather than rational thinking. Moreover, and this is Greene's most important normative point, the epistemic relation between automatic emotions and deontological moral judgment is not a benign one: the former do not facilitate moral insight and they are not, he maintains, conducive to finding moral "truth." In fact, they tend to mask the utilitarian considerations that are relevant to the moral assessment of actions. In a nutshell, Greene's two main claims about moral judgment are that (i) a certain subclass of moral judgments, namely deontological judgments, is based on automatic, intuitive, and emotionally charged processes, and that (ii) these processes pick up on morally irrelevant features of the situation, which generally renders these judgments ill-founded.

Recently, Selim Berker has shown that Greene's argument can be reconstructed in a way that allows for a clear distinction between its empirical and its normative component. Following a simplified version of Berker's (2009) reconstruction, I shall refer to it as the argument from morally irrelevant factors:

Argument from morally irrelevant factors

(1) Deontological intuitions stem from emotional processes.
(2) These emotional processes respond to morally irrelevant factors.

Thus,

(3) Deontological intuitions—and the judgments they give rise to—have no genuine normative force.

Premise (1), which contains the empirical component, holds that it can be shown empirically that deontological judgments are based on emotional reactions toward moral scenarios rather than cognitive processes. Premise (2), which contains the normative component, holds that the situational features that trigger deontological intuitions are morally irrelevant, and that therefore (3), deontological judgments are unjustified. In the following sections, I shall argue that neither of the two premises is tenable; consequently, Greene's conclusion concerning the normative force of deontological judgments is cast into serious doubt.

The following chapter has eight sections. Section 1 gives a brief overview of Joshua Greene's dual process model of moral judgment, and singles out its main claims and predictions. I will merely report what they are, and not assess their validity at this point. Section 2 is about why Greene and others think that this model has significant normative implications. Sections 3 to 5 deal with the first premise of the above argument, and develop a systematic critique of the empirical credentials cited in support of the claim that deontological judgments are based on emotional reactions. I argue that the empirical evidence does not fare well under closer scrutiny, and that the empirical support for the first premise turns out to be weak. Section 6 discusses the second premise, and argues that Greene's argument for the claim that deontological intuitions have no normative force ultimately fails to convince. Just as much as Greene fails to show *empirically* that deontological judgments are driven by emotion, he fails to show *conceptually* that deontological judgments are typically unjustified and that there is an exclusive connection between consequentialist judgment and rational cognition. But even if the first premise in the above argument were true, it would not undermine the normative force of nonconsequentialist assessments of morally significant cases. The empirical evidence and the normative argument for Greene's model do not give us reasons to consider deontological judgments ill-founded. It remains a question of whether one can find evidence undercutting a particular moral intuition, rather than a strike against a whole type of moral judgment. The seventh section presents evidence that there is no link at all between counterintuitive modes of processing and consequentialist judgments. Controlled cognition supports counterintuitive judgments, regardless of whether these judgments are consequentialist or deontological. Finally, I argue that deontological moral judgments are based on educated deontological intuitions (section 8). In particular, I discuss recent evidence that suggests that the acquisition, formation, and maintenance of patterns of deontological moral judgment are based on rational mechanisms of statistical learning.

Deontological intuitions *are* often justified. Note that in making this claim, I do not wish to incur any particular commitment to deontology or Kantian moral theory—or consequentialism, for that matter. The bottom line of this chapter is that the neuroscientific evidence does not support one over the other. I argue that moral judgments, including deontological judgments as a subclass, are based on educated and rationally amenable moral intuitions. These overlearned habitual responses can generally be expected to be reliable and cognitively efficient guides through the real world of moral decisions, a world that is, unlike the hypothetical scenarios Greene uses in his experiments, not inhabited by trolleys, footbridges, unusually large strangers and ticking clocks. Admittedly, our intuitions are prone to misfire. This does not, however, constitute a fundamental strike against the normative force of deontological intuitions, and it does not imply that deontological judgments generally pick up on morally irrelevant factors.

7 Educated Deontological Intuitions

Introduction

There are three basic types of dual process theories: servile, competitive, and cooperative. According to social intuitionism, reasoning is the *slave* of intuition: the latter commands, the former obeys. Greene highlights the *competitive* aspects of his theory. In his version, the emotionally charged, automatic processes and the conscious, cognitive processes compete with one another to generate outputs such as moral judgments. The educated intuitions model belongs to the *cooperative* camp: cooperative models assume that System I and II work together: one provides the intuitive and resource-independent raw material for human cognition; the other monitors the output of its evolutionarily older relative, and either approves of its performance or intervenes by correcting, filtering, or vetoing it. Importantly, Greene doesn't merely say there is a competition; he also says that there is one competitor who typically *deserves* to win. This last claim is the one I am disputing. By pointing out some of the problems with Greene's account, the cooperative profile of my own model will become sharper.

1 Moral Dilemmas and Moral Intuitions

Dual system theories in general are not without problems. As Evans (2008) points out, it is important to clearly identify the characteristics that a process must share in order to belong to one of the two categories. In the case of System II, for example, Evans proposes four such characteristics: the mode of processing (conscious vs. unconscious), its evolutionary age, its functional characteristics (for instance, parallel vs. sequential processing, or domain-specific vs. domain general processing), and its dependence on individual differences (does cognitive performance reflect differences in intelligence or working memory capacity?). However, if one devises such a list of necessary characteristics for one of the systems, and combines this list with the claim that System I and System II cover most, if not all, of the territory of the mental, then all the processes which do not share those features will automatically have to be considered part of the opposite system, simply in virtue of the fact that they do not share the aforementioned features, and regardless of the great disunity there might be between members of this residual group. As a result, moral intuitions are now considered part of the same automatic system that is responsible for perception, acquired habits, or implicit deductive reasoning. In what follows, however, I will mostly ignore this problem. I will assume that dual process theory is a useful framework for thinking about cognition, and leave its internal problems to one side.

Greene is interested in the psychological processes underlying moral judgment, which he studies using neuroimaging techniques as well as behavioral methods. Most famously, he has studied experimental subjects' responses to prevalent moral dilemmas, involving, among others, the so-called *Trolley* and *Footbridge* dilemmas[2] (Foot 1967; Thompson 1976; the exact wording of all dilemmas referred to in this chapter can be found in the notes).[3]

In a series of studies, test subjects were asked to give a moral assessment of the scenarios while undergoing brain scans. What Greene and his colleagues found was that most people tend to say "Yes," it is appropriate to hit the switch, and "No," it is not appropriate to push the large stranger—and this, in their opinion, calls for an explanation. They hypothesized that what makes people approve of killing the one for the purpose of saving the five in the former, but not in the latter case, is the different emotional engagement the two dilemmas elicit. Thus, their prediction was to find brain areas associated with emotion to be more active in subjects when thinking about the second (type of) dilemma.

Let me briefly recap where the connection between types of cognitive processing and types of moral judgment is supposed to come from. Greene has integrated his core claims into a dual process model of moral judgment. Moral judgment, according to his theory, engages two mental subsystems (Evans 2003, 2008): System I is evolutionarily old, operates unconsciously, quickly, effortlessly, with emotional valence ("hot") and without explicit deliberation. System II is evolutionarily recent—indeed, it is often thought to be uniquely human. It draws on scarce cognitive resources and generally works slowly, effortfully, analytically, and in an emotionally cold way (Gilovich, Griffin, and Kahneman 2002; Kahneman 2003). Using the two-systems vocabulary, Greene reframes his explanation of people's differential responses: deontological judgments are based on automatic, emotionally valenced snap judgments that stem from System I, consequentialist judgments are typically based on controlled cognitions that stem from System II.

This difference, Greene maintains, can be mapped onto brain regions that are known to be associated with either the automatic/emotional System I, or the controlled/cognitive System II. What he found was a consistent and significant activation of brain regions—the medial frontal gyrus, the posterior cingulate gyrus, and the angular gyrus—that had previously been established to be associated with emotion (Maddock 1999). Accordingly, Greene concludes that

> the crucial difference between the trolley dilemma and the footbridge dilemma lies in the latter's tendency to engage people's emotions in a way that the former does not. The thought of pushing someone to his death is, we propose, more emotionally salient than the thought of hitting a switch that will cause a trolley to produce similar consequences, and it is this emotional tendency that accounts for people's tendency to treat these cases differently. (Greene et al. 2001, 2106)

As the dual process model holds that moral-personal dilemmas trigger a strong affective response, it was predicted that responses of "appropriate" to this type of dilemma should consume more time. Presumably, subjects would have to overcome their immediate emotional reaction toward pushing the person off the footbridge or smothering a baby to protect a group of people from being found and killed by enemy soldiers, as another scenario has it. Greene's original response time data seemed to confirm this prediction (I shall return to this issue below).

In a 2008 paper, Greene and his colleagues attempted to buttress the response time evidence. They put test subjects under increased cognitive load in order to see if, and to what extent, controlled cognition and emotional intuition are at work in moral judgment tasks.[4] The hypothesis was that if consequentialist judgments primarily draw on cognitive resources, the depletion of those resources should diminish consequentialist responses in some way or another. Interestingly, cognitive load does not increase the overall frequency of deontological judgments (contrary to what had been predicted, because it had been hypothesized that under cognitive load, people have to rely on a quick emotional evaluation of the scenario instead of cool and sober utility calculation). It only increases the *time* people need to give the utilitarian answer they favored anyway.

Patients who suffer from a particular kind of focal brain damage (especially damage to the ventromedial prefrontal cortex [VMPFC]) suffer from impaired emotion (Damasio 1994). The emotional life of these patients is diminished, they are less responsive to cues indicating distress, and they suffer from impoverished social, self- and other regarding emotions like empathy or guilt (Haidt 2003). Michael Koenigs and his colleagues (2007) conducted a study that investigated the performance of VMPFC patients on Greene's moral judgment tasks. It was hypothesized that due to their emotional impairments, individuals with that kind of lesion should be less susceptible to the emotional impact of moral dilemmas like *Footbridge*. It was found that patients are, in fact, more likely to endorse the proposed action of pushing the large stranger as appropriate.[5]

These findings seem to cohere with the main claims of the DP model. The data originally collected by Greene and his colleagues shows that most people are unwilling to approve of performing the respective action in moral-personal dilemmas. Now if that disapproval is due to the emotional revulsion people experience at the very thought of acting in the described way, then people who suffer from flattened and less differentiated emotion should have less to struggle with.

2 "From Neural 'Is' to Moral 'Ought'"

Greene has not restricted his account to the descriptive task of finding out what the (neural) processes that underlie moral judgments are. Indeed, he has explicitly stated that his theory provides the building blocks of a "debunking explanation of moral realism" (Greene 2003, 849). He argues that, based on the evidence about the typical causal genesis of our deontological judgments, we can tell a story that undermines their foundation as a whole. If it were true that deontological judgments do not stem from pure reason, but are actually triggered by the "up close and personal" features of morally significant cases, this would give us reason to assume that these judgments do not pick up on features of the world that are "out there," objectively and mind-independently, but on internal emotional patterns we have been equipped with by evolutionary mechanisms (for example, natural selection). These patterns can be expected to have evolved in small-scale societies, whose members had no chance to have encounters with each other but in an up close and personal way. Therefore, it is no surprise that our deontological intuitions turn out to be particularly unsuitable when it comes to dealing with cases ranging from giving to famine relief in foreign countries to the permissibility of abortion: our ancestors never confronted such cases (Singer 2005). This, Greene argues, gives us reason to discount our deontological intuitions in favor of our consequentialist ones: the latter are readily applicable to the modern world we happen to live in, whereas the former typically misfire in a way that, upon reflection, we might not be willing to endorse.

3 Functional Neuroimaging and the Problem of Reverse Inference

Greene's findings that deontological judgments engage specific brain regions, his claim that these areas have been found to be associated with emotion, and the implications of these findings are often taken for granted. Criticism of the DP model, for the most part, does not address the methodological question whether evidence *of that type* can support a conclusion like the one drawn by Greene, focusing on conceptual problems instead.

Remember that Greene's claim that deontological judgments are driven by morally irrelevant emotional reactions is based on the evidence that a certain brain area, *A*, has been found to be activated when subjects make deontological judgments. Prior investigations, in turn, have shown that region *A* is involved while performing an emotionally engaging task *T*. This task can consist in looking at emotionally charged facial expressions, or reading sentences with emotional content. Greene's conclusions, then, are drawn on the basis of these previous studies: if *A* is involved while performing *T*, and *T* is of kind *K*, then other mental processes that engage region *A* have to be of kind *K*, too. For example, studies might have shown that whenever subjects look at frightening images, a certain brain region "lights up." This suggests that this region is implicated in the processing of fear. Then one might be tempted to conclude, *independent* from direct knowledge about the nature of a second task, that because the same region lights up when subjects are performing this second task, it must be accompanied by feelings of fear as well.

The DP model relies on the assumption that, from the fact that certain parts of the brain (here: the medial frontal gyrus, the posterior cingulate gyrus, and the angular gyrus), which are involved in the processing of emotionally laden information, are also involved in making deontological judgments, it follows that deontological judgments are based on emotion, too. However, this move is not as innocent as it may seem, and it is far from convincing.

The inferential pattern that a great amount of practice in cognitive neuroscience is based on has this form, call it the forward inference:

Forward inference

(1) Whenever cognitive task *T* is performed, brain area *A* is active.
Thus,
(2) *A* is involved in the execution of *T*.

Forward inferences provide us with correlational data and help to localize functions. However, Greene's claim that deontological judgments are driven by emotional processes is based on an entirely different type of inference, which, following Russell Poldrack (2004, 2006), I shall refer to as reverse inference:

Reverse inference

(1) Whenever cognitive task *T* is performed, brain area *A* is active.
(2) Other studies have shown that whenever cognitive task *T** is performed, brain area *A* is active.
(3) *T** is of kind *K*.
Thus,
(4) Activation of *A* while performing *T* demonstrates that *T* is of kind *K*.

Greene's empirical findings suggest that whenever people make deontological judgments (*T*), increased activity in the aforementioned brain regions (*A*) can be observed (1). He then goes on to refer to previous studies that putatively have shown that the same regions are active (2) when emotionally engaging tasks (*T**) are executed (3). From this, he leaps to the conclusion that deontological judgments must be emotional in nature, too.

There are, however, some important limitations to the reverse inference approach. Generally, it is thought to be tricky to reason from brain structure to cognitive function (Henson 2005; see also Aguirre 2003; Miller 2008; Logothetis 2008). The strength of a reverse inference varies as a function of the selectivity with which the region is involved in the process:

> The greatest determinant of the strength of a reverse inference is the degree to which the region of interest is selectively activated by the cognitive process of interest. If a region is activated by a large number of cognitive processes, then activation in that region provides relatively weak evidence of the engagement of the cognitive process; conversely, if the region is activated relatively selectively by the specific process of interest, then one can infer with substantial confidence that the process is engaged given activation in the region. (Poldrack 2006, 60)

Ideally, the evidence would support a biconditional: it should hold that if, and only if, a subject is engaged in a task of a certain kind, the respective brain region is activated.

But this is hardly ever the case, and the meta-analysis concerning the function of the retrosplenial cortex (Maddock 1999) Greene relies on in making his claims was called into question right away. Brent Vogt and John Absher argued that although the location of the retrosplenial cortex can be identified, "nothing is known about its function. Although there might be a rationale to consider if areas 23 and 31 are involved in emotion, no study has shown a specific involvement" (2000, 196).

A recent meta-analysis (Kober et al. 2008) of which brain areas are involved in emotional processing has confirmed Vogt and Absher's worry, and either mentioned the brain areas Greene's claims are based on only very cautiously, or not at all: "Functionally speaking, PCC [posterior cingulate cortex] seems to play a nonspecific relay role between other emotional-related networks, instead subserving a number of attentional and perspective-taking functions" (Klein 2010, 147).

Complementarily, Greene his colleagues (2004) claimed that cognitive regions such as the dorsolateral prefrontal cortex (dlPFC) show increased activation when subjects judge that the action in a personal moral dilemma is appropriate. The nonspecificity of reverse inferences applies to this portion of the brain as well. But even if it were legitimate to draw a reverse inference from dlPFC activation to cognitive function, it would not support the claim that the deontological/consequentialist distinction can be mapped on the emotional/cognitive distinction. The primary function of the dlPFC seems to lie in working memory (Wager and Smith 2003): an increased activation in that area does not directly bear on whether a mental function is more cognitive, in the sense of rational, but on whether it manipulates a greater amount of information. Increased working memory activations indicate a higher amount of conflict, rather than a higher amount of rational cognition. A mental process can be cognitive even if it deals with only one tiny piece of information and does so very quickly. (I shall return to this point below.)

At any rate, further empirical evidence is needed to strengthen the inference from the activation of a brain region to the involvement of a cognitive process (cf. Machery 2014). The two most important ways to do so are to (i) provide supporting behavioral evidence and (ii) increase the selectivity of response in the brain region of interest, which can be done by showing that damage to the region results in impairment of the function. As mentioned above, Greene has attempted to pursue both strategies by investigating the reaction time people need when making different types of moral judgments, and by providing evidence for the DP model from focal brain damage. I shall discuss these attempts one at a time.

4 Response Time and Cognitive Load

The DP model holds that deontological judgments are triggered by emotion and that consequentialist judgments are based on rational cognition. Presumably, the emotional reactions elicited by moral-personal dilemmas like *Footbridge* are the same for the vast majority of people. Why is it, then, that some people endorse the utilitarian action, although their feelings tell them otherwise?

Greene makes the following suggestion: some people, he claims, override their immediate emotional response in favor of a rational cost-benefit analysis (Greene 2008). Clearly, overriding an emotionally charged intuition consumes considerable cognitive resources, and that, in turn, takes time. We have seen that Greene's findings confirmed this theory, as he found that consequentialist responses of "appropriate" to personal-moral dilemmas took people 6.8 seconds on average, whereas deontological responses of "inappropriate" took 5.0 seconds on average. Greene arrived at this result by calculating the mean for the appropriate responses, and comparing it to the mean of the inappropriate responses (that is, he compared the total sum of time for responses of inappropriate, divided it by the number of responses, and compared that result to the total sum of time for responses of appropriate, also divided by number of responses). This method, however, is liable to systematic errors. For that reason, Berker (2009) has suggested a different method: calculate the average response time of inappropriate answers made by participants *for each question,* and then calculate the mean of those means and compare it to the mean of the means of appropriate answers for each question. Such a procedure, he argues, makes it easier to spot and avoid outliers that spoil the significance of the overall result.

This criticism might strike some as nitpicking. But, as McGuire and his colleagues (2009) have shown, it is not. Indeed, it is a fatal blow against one extremely important piece of behavioral evidence there is for the DP model. There are two problems with the stories Greene based his statistical analysis on: first, in the *Hired Rapist* scenario, there is neither a consequentialist nor a deontological rationale for the action at issue.[6] Accordingly, responses to this dilemma do not bear on the question of which mental processes these types of judgments are based on. (This holds for other dilemmas as well; see Kahane and Shackel 2010.) But another statistically more damaging problem is that the answer is, at least to most people, *completely obvious.* People will say that it is inappropriate to hire the rapist, and, more importantly, they will do so very quickly. Given that Greene had decided upfront to treat responses of "inappropriate" to personal dilemmas as deontological, the results become severely skewed. How does the data change, statistically, when dilemmas with obvious answers and no genuine conflict between utilitarian and deontological considerations are left out? When McGuire and his colleagues reanalyzed the data with dilemmas in which the proposed action was endorsed by less than 5 percent of the people taken out, the RT effect disappeared. Instead, what they found was that RT differences obtain not between deontological and utilitarian responses to the dilemmas, but between moral and impersonal dilemmas in general. Overall, it takes people longer to think about and give a response to a personal dilemma than to an impersonal dilemma. This should not, however, come as a surprise: no one ever denied that personal-moral dilemmas are more conflicting than others (Schaich Borg et al. 2006; Dean 2010). That's what they are designed to be, after all.

Greene has admitted that the interpretation of his original study is flawed, and retracted claims that were based on the skewed data (Greene 2009). The RT effects found in his study were largely due to the inclusion of several "dilemmas" that didn't adequately pit utilitarian and deontological considerations against each other, and, in many cases, weren't even dilemmas to begin with, because the answer was obvious. In the subset of dilemmas (like *Trolley* and *Footbridge*) that do contain a genuine conflict between consequentialist and deontological intuitions, there is *no RT effect*.

As a reaction to this problem, Greene designed a follow-up study in which people were asked to perform his moral judgment tasks under increased cognitive load. The prediction was that cognitive load should affect either RT, or frequency of utilitarian judgment, or both. The results, however, did not confirm the dual process model.[7] It seems that the RT evidence, as it stands now, does not force upon us an interpretation according to which it is the automatic emotional response that is triggered by the "up close and personal" nature of some high-conflict dilemmas that makes people make deontological judgments. Neither does it suggest that consequentialist judgments depend on controlled cognition. If it were true that the two types of judgments could be mapped onto different mental subsystems, that should be invoked in subjects in general whenever their "emotional buttons" are pressed in the same way. But the finding that there are "high" and "low" utilitarian subjects suggests that some people are, and some aren't, already equipped with a utilitarian tendency, a bias that figures in their patterns of moral judgment anyway. The evidence suggests that which types of judgment subjects favor is not so much a question of whether they rely on emotional or cognitive processes in making them, but of whether they are prone to be swayed by certain moral considerations. People's intuitions have already been shaped by prior moral reasoning and by habitualized intuitive responses. This has a greater impact on their behavior than the features of the particular scenario they are asked to assess. Evidence about the impact of different—controlled or intuitive—cognitive styles on people's moral judgment confirms that conjecture (Bartels 2008; Moore, Clark, and Kane 2008). Subjects who place more weight on consequences rather than rules in their thinking *in general* favor consequentialist or deontological judgments, respectively.

This means that another central empirical prediction of the DP model has not been borne out. Indeed, it has repeatedly been shown to be empirically untenable. It is not the difference between controlled cognition and automatic emotion that explains why some people prefer either the utilitarian answer or the deontological one, but the fact that people have acquired different moral views, and assess the dilemmas presented to them accordingly. That implies that both types of judgment are cognitive and automatic at the same time: they are cognitive in virtue of the fact that they are based on acquired, rule-based theories about what is right and what is wrong, and they are automatic to the extent that these acquired, implicit theories automatically and intuitively guide people's answers, and generate certain types of judgments. Returning to the last section, we have to conclude that the behavioral data does not strengthen Greene's reverse inference from the activation of certain brain regions to the nature of the cognitive processes involved in the tasks.

5 Brain Lesions and the Dual Process Model

The second type of empirical evidence that is supposed to provide further support for Greene's reverse inference from the involvement of certain brain areas to the emotional nature of deontological judgment comes from research studying the effects of brain damage on the performance in moral judgment tasks. Greene explains the "utilitarian" performance of VMPFC-damaged patients with their emotional deficits.

One minor empirical problem seems to be that, as observed by Moll and de Oliveira-Souza (2007), the patients who took part in Koenigs and his colleagues' (2007) study also had damage to a region (the dorsolateral prefrontal cortex) that Greene holds to be responsible for paradigmatically consequentialist judgments. If one takes Greene's reverse inference seriously, these patients do not provide clear support for the DP model. (See, however, Mendez, Anderson, and Shapira 2005 and Ciaramelli et al. 2007 for experimental evidence for the DP model that overcomes that problem.) The argument developed in this chapter remains unaffected by that point: activation of the dlPFC only bears on the nature of the cognitive tasks that are at issue here if a reverse inference from brain region to function can be validated, a point I have already challenged above.

One can argue that lesion studies are not apt to support the claim that utilitarian judgments are associated with rational cognition from yet another angle. In a different paper, Koenigs and Tranel have shown that the judgmental and behavioral patterns of patients with prefrontal damage often cannot be attributed to the absence of emotions, but to an increased presence of different emotions: the poor performance of VMPFC patients in the Ultimatum Game, Koenigs and Tranel propose, is due to the fact that patients with "VMPC damage tend to exhibit exaggerated anger, irritability, emotional outbursts, and tantrums, particularly in social situations involving frustration or provocation. ... All seven VMPC patients who participated in this study have demonstrated such behavior in their personal lives" (Koenigs and Tranel 2007, 954). It is difficult to see why the willingness to endorse the consequentialist option of pushing the fat man in *Footbridge* should not be attributed to this heightened impulsivity, aggressiveness, and diminished behavioral control.

Elsewhere, Greene cites evidence from mood induction in defense of his model (Greene 2009, 581–584). Valdesolo and DeSteno (2006) have found that inducing positive emotion (by making people watch 5 minutes of comedy, for instance) elicits more utilitarian judgments: people are more willing to sacrifice the large stranger when they are in a better mood. Their patterns of judgment resemble those of the VMPFC patients. This evidence does not at all support Greene's theory, however: the evidence from mood induction might well be taken as an empirical strike *against* the claim that consequentialist judgments are driven by cognition rather than emotion. They might simply be driven by emotion from the other side of the spectrum. They are not, one could say, driven by a negative response toward killing a person, but by an unusual lightheartedness, good mood, or an unusually high degree of happiness. It is just that different emotional reactions are involved, rather than none at all.

The evidence does not support the normative point that consequentialist judgments can be trusted whereas deontological intuitions ought to be discounted. It has been shown, after all, that a good mood can enable people to perform wicked actions more easily. To use a strong example, we can think of Zygmunt Bauman's (1982) observation that bureaucratic organization and division of labor helped conceal the emotionally engaging aspects of genocide during the Holocaust. Greene's argument can be turned upside down: instead of arguing that deontological intuitions should be discounted because they are elicited by the emotionally salient features of a situation, one could say that consequentialist intuitions should be discounted because the opposite is true for them: they respond to emotionally *unengaging* features of a situation, or to put it differently, they are triggered by scenarios whose emotional significance can more easily be ignored.

The fact that patients with focal damage to the VMPFC make more utilitarian moral judgments undermines the normative point Greene and, following him, Peter Singer (2005) are making. In most other cases, especially *in their own real lives* outside the lab, these patients are known to be dramatically poor moral and practical reasoners. The type of utilitarian judgment Greene and Singer claim to be normatively superior is typically made by people whose judgments are known to be normatively inferior. In the end, the DP model faces a dilemma: either the undisputedly irrational behavior exhibited by VMPFC patients counts as evidence that the VMPFC is not an area that is responsible for mere emotional processing—because if emotion and reason are distinct, why does damage to an emotional area result in irrational behavior?—or it counts against the idea that the moral judgments those patients are inclined to make ought to be normatively privileged (Maibom 2005).

6 Up Close and Personal? From the Personal/Impersonal Distinction to the Concept of "Personal Force"

The second premise in the *Argument from Morally Irrelevant Factors* contains two different theses. To make his normative point, Greene has to show that, first, subjects really are responding to the very factors he deems morally irrelevant (an empirical point within the normative point) and that, second, the factors he deems morally irrelevant and that subjects allegedly are responding to really are morally irrelevant. Let me address both issues one at a time.

If we look at the differences between the *Footbridge* and the *Trolley* dilemma, Greene maintains, we only see differences in morally irrelevant factors. Therefore, we have reason to assume that these morally irrelevant differences explain people's differential responses, and that we ought to discount the intuitions that give rise to them. Morally irrelevant factors do not convey any justificatory force on the deontological judgments they elicit, which undermines this class of judgments as a whole. But what exactly is the factor that explains participants' differential responses? Originally, Greene held that it is the "personalness" of a scenario that makes people disapprove of an otherwise consequentialistically justified action.

Let me briefly explain why the DP model in its current from does not make that claim anymore.

For the sake of the argument, I will grant that features that make a moral scenario personal or impersonal are morally irrelevant. Why is that supposed to be the case? Here, we can make a short detour through Peter Singer's (1972) famous argument: if we compare the "child drowning in a shallow pond before our eyes" scenario with the "child starving to death in a foreign country" scenario, it seems natural to assume that what triggers our inclination to help in the first case but not in the second is mere spatial distance, immediacy of experience, and the resulting vividness of being directly confronted with one human being's suffering, but not the other's. (One could say that moral properties do not supervene on spatial properties.) Singer offers an evolutionary explanation for why our moral intuitions misfire so severely in his example: roughly, he holds that our moral intuitions are tailored to small-scale societies, and in our phylogenetic past, we didn't encounter situations that called for altruistic behavior over long distances. Instead, we have developed immediate emotional reactions that regulate our behavior when it comes to close interactions. We simply have not been selected for our moral sensibility toward the suffering of complete strangers on the other side of the globe (Singer 2005). But this means that our emotional responses have not been selected to track the moral truth: "what is the moral salience of the fact that I have killed someone in a way that was possible a million years ago, rather than in a way that became possible only two hundred years ago?" (2005, 348). We might think that stabbing one's wife to death with a knife is far more cruel and violent than killing one's husband by poisoning him, but it seems obvious that both actions are equally ruthless, and equally morally wrong.

Greene used to favor a similar explanation for why people respond differently to *Trolley* and *Footbridge*. The only difference between the two, he argues in his 2001 and 2004 papers, is that in *Trolley*, the killing is done using a lever, whereas in *Footbridge*, the killing is done in an "up close and personal" way. He maintains that actions which have a "me hurt you" structure (that is, a structure in which *physical harm* is inflicted by *me* onto a *concrete other person*) are generally thought to be less permissible than actions where harm is brought about in a more indirect "impersonal way." But whether something—the *same* thing, indeed—is done in a way that involves physical contact or not does not seem to bear on that something's moral right- or wrongness. Moral properties do not supervene on the personal/impersonal distinction, which renders the distinction morally extraneous.

Marc Hauser and his colleagues designed scenarios that are slightly different from the original *Trolley* and *Footbridge* cases. They were particularly interested in whether people rely on the doctrine of double effect in making their judgments, and altered the scenarios presented to test subjects accordingly, in order to exclude the possibility that differences in subjects' judgments were due to the difference between (a) the redirection of an existing and the introduction of a new threat or (b) the personal- or impersonalness in which harm is brought about, as Greene would propose.[8]

What Hauser and his colleagues found was that people do see a morally relevant difference between *Heavy Object* and *Trolley*. This suggests that, contrary to the DP model, the personal/impersonal distinction cannot be mapped onto the deontological/ consequentialist distinction (Kamm 2009). The "me hurt you" criterion that distinguishes personal from impersonal dilemmas does not apply to *Heavy Object*, yet a significantly higher amount of test subjects (*Heavy Object*: 44 percent vs. the standard trolley problem: 15 percent) judged the action morally impermissible in this scenario. Subjects do not primarily respond to "personal" factors in making deontological judgments. Rather, they seem to disapprove of using a person as a mere means to an end, which is done in both *Footbridge* and *Heavy Object.*

Hauser's findings, among other things, eventually forced Greene to give up the personal/impersonal distinction and replace it with a more sophisticated proposal:[9] he now maintains that people's differential responses to the dilemmas are due to a factor he calls "personal force": "An agent applies personal force to another when the force that *directly* impacts the other is generated by the agent's muscles, as when one pushes another with one's hands or with a rigid object" (Greene et al. 2009, 365). I will now examine the empirical evidence that motivates this proposal and show that it is both empirically implausible and does not help to support the core normative point in the DP model, namely premise (2) in the *Argument from Morally Irrelevant Factors*.

Greene and his colleagues (2009) designed additional scenarios that factor in differences in intentionality and relative spatial proximity. They used four different versions of the footbridge case, including the original one, two slightly altered versions of the trapdoor variation of the footbridge case (which are similar to *Heavy Object* in the most important respect, namely that one person is impersonally used as a means to stop the trolley), and a newly constructed one in which the agent can push the victim with a pole.[10] Hauser's *Heavy Object* variation of the trapdoor case (in which the person that is used as a means to stop the trolley stands on a side track instead of being dropped onto the tracks through a trapdoor) is an attempt to bring out the significance of the means/side effect distinction and challenge the importance of the original personal/impersonal distinction. Greene's more recent findings, however, show that although there is no significant difference in people's responses between *Footbridge* and *Footbridge Pole*, there are significant differences between these two and the two that bear important similarities to *Heavy Object*.[11] The (unconscious or conscious) employment of the principle of double effect—a good candidate for a morally *relevant* consideration—cannot explain this finding, but the PF factor—a good candidate for a morally *irrelevant* consideration—can. *Footbridge Pole* and the trapdoor/*Heavy Object* scenario are very similar: the proposed action in neither of the two involves direct physical contact and in both cases, the victim is used as a means to stop the runaway trolley, yet people are twice as likely to find it impermissible to push the person using a pole than to drop the victim onto the tracks or to divert the trolley onto a sidetrack where it will then be stopped by the victim. This significant difference within the spectrum of deontological responses is left unexplained by the principle of double effect. But it is explained by the PF factor, because only in *Footbridge Pole* does the agent affect the victim in a way that is directly mediated by the agent's muscular force. But surely, this can have no bearing whatsoever on which proposed course of action is morally appropriate or not.

There are in turn further empirical findings that are left unexplained by the PF factor. Greene states that personal force is exerted from one agent to another. Why is it, then, that people show the same differential response when moral dilemmas are about teacups rather than people (Nichols and Mallon 2006)? Nichols and Mallon's findings suggest that moral judgments critically depend on the (possibly implicit and unconscious) application of rules; this happens, they argue, in a way that rules out *any possible* kind of "personal" factor suitable to fully explain people's differential responses. In their teacup cases,[12] two of Greene's criteria for personal force do not obtain:

there is no agent that does anything to another agent. (There is only one agent that does something to a teacup.) And second, there is no muscular force involved that directly impacts the teacup, because one teacup is *thrown* on another. As far as empirical support is concerned, the PF factor and Hauser's as well as Nichols and Mallon's rule based accounts are on an equal footing: each of the proposals leave some of subjects' responses unexplained.

What about the normative force of the PF factor? Telling a story about the evolutionary genesis of moral intuitions is a major part of Greene's and Singer's debunking explanation of the appeal of deontological considerations. But with the concept of personal force, this evolutionary story drops out of the argument. It is far from clear why human minds would have been selected for being sensitive to something as specific as muscular force. If automatic responses reflect their evolutionary origins, it is plausible to assume that, given the circumstances under which we used to live, evolution could have equipped humans with an aversion toward up close and personal harm. But the PF factor, as it is defined by Greene, seems too specific to be a good candidate for an environmental feature that human beings have evolved to be sensitive to. One could have argued that the proper function of deontological intuitions is to detect up close and personal harm, but it would be arbitrary to maintain that they are established to be set off by only those actions that comprise a PF factor. It would not be an evolutionary advantage to be sensitive to harm that is brought about in a way that involves an agent's direct muscular force any more than to, for instance, the throwing of an object. But the evolutionary debunking of deontological intuitions is something the DP model cannot easily do without. One of the main reasons Greene gives us for thinking that deontological intuitions are unreliable is that these intuitions are tailored to conditions in which we no longer find ourselves. But it is hard to see why the PF factor should have mattered so much to our ancestors and, hence, why our responses to that factor should be tailored to these outdated conditions in the first place.

It is correct that the means/side effects distinction—and thus the doctrine of double effect—cannot explain why people deem *Footbridge Pole* to be worse than, say, *Heavy Object*. But the principle of double effect is not the only alternative explanation for this finding. Rather, if it is true that people are sensitive to actions that involve, among other things, a PF factor, it may very well be that this is not because people's minds specifically pick up on that very factor, but because actions with features that typically are morally relevant, such as aggressive acts or acts that involve the specific intention to actively harm somebody, are often instantiated by actions that also involve personal force as one of their elements. It is tacitly assumed by Greene that these things cannot matter to moral evaluation. He restricts the realm of possible morally relevant factors to consequences. But if that is the case, then his argument does not *support* consequentialism anymore, but *presupposes* its truth. The range of things that are candidates for proper moral assessment go beyond (actual, expected, or intended) consequences in the directly observable world: we take into consideration as morally relevant an agent's intentions, motives, and character traits all the time, and one can plausibly make the case that the disposition to not mind directly harming another person or the willingness to apply—without further ado or scruple—personal force to another human being are tendencies that typically are manifestations of certain vicious traits, and that that is why people disapprove of those kinds of actions. Again, this is not to say that the deontological response to the discussed moral dilemmas is the *correct* one. It only means that the factor deontological intuitions pick up on is not necessarily *morally irrelevant*.

Regarding a moral theory's *overall* sensitivity to morally relevant factors, one could follow Kamm (2009) in arguing that the version of act-consequentialism Greene relies upon is far too crude to account for a variety of morally relevant considerations, which it would therefore be likely to overlook. The so-called *Crying Baby* case, for example, describes a situation in which the only way for a group of civilians hidden in a cellar to escape being discovered and killed by the wartime enemy is to smother a crying baby. Act-consequentialism holds that it would be permissible to do so—in fact, it is committed to there being an *obligation* to kill the baby. But this theory runs roughshod over the (morally relevant) differences between this case and ones in which, for instance, the death of the baby would not also be unavoidable or in which one could use the baby's blood to save two soldiers from bleeding to death. This suggests that it is very much an open question which of the two theories is normatively preferable when both *sensitivity to irrelevant* as well as *insensitivity to relevant* situational features are tallied.

7 Intuitive and Counterintuitive Moral Judgment

Greene's debunking argument against deontological intuitions relies heavily on the idea that such intuitions are produced by cognitive processes that are obsolete: they have evolved under conditions we no longer inhabit, which explains both why they are intuitive and likely to be untrustworthy. And the basic idea behind this type of debunking is indeed correct, as many players who have practiced their service on clay, but now want to try their luck on grass will be able to confirm.

There are several types of evidence used to support this picture, many of which I have discussed and criticized at length in this chapter. But notice that if it turned out that, at the most basic level, the very association of deontological with intuitive (and thus presumably outdated and unreliable) and consequentialist with counterintuitive (and thus hopefully more reliable) cognitive processes did not work, Greene's debunking argument could not *possibly* go through.

Yet this is exactly what has been found.[13] To see whether there is any intrinsic relationship between deontological and intuitive judgments on the one hand, and consequentialist and counterintuitive judgments on the other, Kahane and his colleagues (2012) designed vignettes where the link between process (intuitive vs. counterintuitive) and content (deontological vs. utilitarian) is reversed. Scenarios where the deontological option is the counterintuitive one involve, for instance, several white lie cases and many others in which a greater harm conflicts with some more or less strict prohibition to harm. One of them reads like this:

> You are a waiter. You overhear one of your customers say that he is about to go to jail and that in his last forty-eight hours of freedom he plans to infect as many people as possible with HIV. You know him well enough to know that he is telling the truth and that he has access to many potential victims. You happen to know that he has a very strong allergy to poppy seeds. If he eats even one he will go into convulsions and have to be hospitalized for at least forty-eight hours. Should you cause this man to have a serious allergy attack in order to prevent him from spreading HIV? (Kahane et al. 2012; see Supplementary Data, 4)

The data suggest that counterintuitive judgments are perceived as more difficult than intuitive judgments, regardless of their content. It took subjects longer to arrive at deontological judgments than utilitarian ones in cases where the utilitarian option was designed to be the intuitive one. This, in turn, suggests that Greene and his colleagues found consequentialist judgment to be associated with controlled cognition because in the dilemmas they used, the consequentialist response always happened to be the counterintuitive one.

To be fair, Kahane's results have been called in question (Paxton, Bruni, and Greene 2013) on empirical grounds; the current state of the debate seems to be a draw. But this much is obvious: counterintuitive deontological judgments exist. I see no reason why these judgments should not be supported by roughly the same cognitive processes (and corresponding brain regions) which are in charge of controlled cognition in general. Greene's indictment of deontological judgments thus fails to the extent to which it is based on their intuitive compellingness.

Moreover, just like deontological judgments do not seem to be as closely connected to intuitive processes as originally supposed, utilitarian judgments are unlikely to be generated by *reasoning processes* which deserve to be described as utilitarian in nature. Note that utilitarian moral reasoning is defined not as reasoning that takes utilitarian, harm-centered considerations into account, but reasoning that *only* takes such considerations into account. All other forms of reasoning, even if they lead to judgments that coincide with the ones utilitarianism would recommend, are nonutilitarian in nature. Now consider the following sequence:

(1) We have reason to reduce harm (or prima facie *Duty to Save*). ...
(2) 5 lives > 1 life.
(3) There is reason not to intentionally harm an innocent person (or prima facie *Duty Not to Harm*).
(4) In *this context*, there is *more* reason to minimize harm.
(5) It is permissible to sacrifice 1 to save 5. (Kahane 2012, 531–532)

Kahane argues that it is far more plausible to suggest that this is what goes on when people arrive at a moral verdict that also happens to be favored by utilitarian moral theory, and recent evidence from conflict detection seems to confirm this (Białek and De Neys 2016). However, the above inference exhibits a pattern of characteristically *deontological* reasoning. Worse still, utilitarian judgments do not seem to be associated—as we would expect them to be—with impartial concern for the greater good, but with far more sinister dispositions (Kahane et al. 2015).[14] People who are more likely to endorse the utilitarian option in *Trolley*-style scenarios are less likely to endorse charitable giving and other altruistic concerns, and more likely to endorse downright immoral actions in business and other contexts.

Obviously, none of the findings I have mentioned are unambiguous and none of the interpretations I have suggested are without alternative. But that simply seems to be the way it is: moral judgments stem from different, potentially conflicting and potentially converging mental operations, and are sensitive to different types of moral considerations, including cost-benefit analyses as described by consequentialism, the application of moral principles as described by deontological theories, the evaluation of motives and traits as described by virtue ethical approaches, emotional reactions which underpin all of these and—which I have not denied at any point in this book—many biases, ill-founded assumptions, and other distorting influences.

8 Educated Deontological Intuitions

Where does all this leave the DP model? Maybe the criticism it has received refutes the usefulness of the personal/impersonal distinction and the PF factor, but leaves the core claims of the DP model—most importantly, that different types of moral judgment are performed by different mental subsystems—untouched. As we have seen, this does not seem to be the case: the normative part of Greene's argument is built almost entirely upon claims about factors people pick up on in making their judgments. It does not matter for this part whether these are described in terms of the personal/impersonal distinction or the PF factor. Greene claims that deontological intuitions have no genuine normative force because they pick up on features that are morally irrelevant. And they are supposed to be morally irrelevant because they are located on a level that does not bear on the moral rightness and wrongness of actions. While this last point might be correct, it does not do any work in the argument, because the available proposals for what influences people's judgments do not adequately explain these responses. If the claims about morally irrelevant factors are dropped, however, the whole normative point has to be dropped, too; and if all that remains from the DP account is the automatic/controlled distinction, then its normative part collapses as a whole. After all, it is not clear why the fact that a (type of) moral judgment is arrived at automatically should undermine its normative force. Moreover, the finding that some moral judgments are executed by automatically operating processes and some by processes that draw on controlled cognition is not an interesting new result, but a *truism*.

Another option for Greene might be to argue that consequentialist judgments are justified simply because they are driven by controlled cognition *and that very fact* renders them trustworthy. But in making such an argument, he would be mistaken about the conceptual connection between a process's being *cognitive*—in a naturalistic sense—and its being *rational*—in a normative sense. Obviously, whether a cognitive process consumes more time and resources or whether it is diminished by increased cognitive load does not bear on the rationality of that process, or the rational quality of the judgment that is being arrived at, but merely on the *amount of information* that is being processed. It will most certainly take subjects longer to calculate 563 + 789 than 5 + 7, but that doesn't mean that the former "mathematical judgment" is more and the latter less rational, and it certainly doesn't undermine the truth of 5 + 7 = 12. More complex calculations simply consume more working memory and rely on a greater amount of controlled cognition. This does not, however, support a dual process theory of mental arithmetic. Examples of mental processes that are *automatic yet rational* are legion: every day, people avoid hitting animals with their car by quickly changing traffic lanes without any conscious deliberation because they have been trained to do so, and experts rapidly distinguish between fake and original antique furniture. (In fact, it is the very definition of expertise to have acquired the capacity to make reliable snap judgments in a certain domain.) Whether these judgments have normative force, in turn, depends on whether they stem from a reliable process of belief formation. It does not depend on how quickly they are made, whether the rules they are based on are consciously accessible, or which brain area they are performed in.

The normative premise in the argument from morally irrelevant factors lacks empirical support: in making deontological judgments, subjects are not responding to the "up close and personal" factors Greene considers morally extraneous. This does not yet show, however, that deontological intuitions are justified. What do deontological intuitions pick up on? And do the factors they pick up on convey any normative force on those intuitions? I have shown that what is left of the DP model is that moral judgments are sometimes made in an automatic and sometimes in a cognitively controlled way, *but that fact* does not bear on the question of which moral judgments are or are not justified. For the most part, I have argued that deontological judgments are *not unjustified*, but did not attempt to show that they are justified. Let me now sketch some possibilities as to how deontological intuitions might be vindicated.[15]

The easiest strategy might seem to be to charge Greene's and Singer's normative argument of committing the naturalistic fallacy. Obviously, one cannot directly derive an "ought"—that we ought to discount our deontological intuitions—from an "is"—because the patterns that explain those intuitions have been shaped by natural selection in our evolutionary past. I do not think, however, that this strategy is successful. Singer and Greene do not directly derive an ought from an is; what their argument tries to achieve is to shift the burden of proof on those who defend deontological judgments. It seems that beside their intuitive appeal, not much can be said in their favor. But if their intuitive appeal can be debunked, which Greene and Singer try to do, what else does one have to go on?

Here is another strategy. Moral judgments, I have argued, are based on educated intuitions: overlearned, habitualized judgmental responses to morally salient scenarios. This strategy has two advantages: first, it shows why it is justified to trust our moral intuitions (including our deontological ones) in most *ordinary* cases. Second, it allows for our moral intuitions to misfire in *deviant* cases, and explains this misfiring in terms of the limited capacities to process information finite beings like us have at their disposal (Klingberg 2009). The view is attractive for those who argue in favor as well as those who argue against the normative force of deontological intuitions. It can account for the possibility that the consequentialist response to *Footbridge* might, all things considered, be the better response, but leaves that question open to further normative argument.

Note that the claim that deontological judgments are based on educated intuitions—which are mental, but not necessarily *emotional* operations—might count as a denial of both premises in the argument from morally irrelevant factors. I have dealt with the first one extensively already, and shall focus here on the second premise.

There is a problem with the way the DP model phrases its main argument that is located on a deeper, conceptual level. From the very start, the model tends to conflate three distinctions: the distinction between *automatic* and *controlled* processes; the distinction between *emotional* and *cognitive* processes; and the distinction between *justified* and *unjustified*, or reliable and unreliable, processes of judgment formation. One mistake that many critics of the DP model have made is to buy into this conflation.[16] This has motivated them to argue that deontological judgments are not based on emotional processes, and that they are, in fact, much more cognitive than the evidence suggests. But in doing so, they have engaged in a wrong, or at least a misguided, battle.

The alternative is to reject this conceptual confusion. Automatic processes need not be emotional at all: nonemotional, but nevertheless quick and effortless intuitions in logic, language, or physics are apt counterexamples to this assumption. And automatic processes need not be unjustified, either. Judgmental heuristics are a particularly good example of mental processes that resist the conflation of the *automatic* and *controlled* distinction with the distinction between *rational* and *nonrational* mental processes. The intuitions that spring from these rules of thumb are susceptible to sophisticated forms of education. I will return to this point later.

The real world is full of computationally intractable problems which controlled cognition is ill-equipped to deal with. This has important methodological implications for the empirical study of moral judgment. Gigerenzer, for instance, recommends that we

> study moral intuitions in natural environments, or in experimental models thereof … rather than using hypothetical problems only. Toy problems such as the "trolley problem" eliminate characteristic features of natural environments, such as uncertainty about the full set of possible actions and their consequences, and do not allow the search for more information and alternative courses of action. (2008, 11)

The focus on real world situations rather than artificial scenarios not only provides a hint as to why many of our educated moral intuitions are justified and reliable, but also why deontological intuitions sometimes appear to lead us astray. The concept of a "moral emergency" is crucial here (Appiah 2008, 98). Scenarios like *Footbridge* are defined by four features: (i) A decision has to be made in a very short period of time. (ii) There is a clear and simple set of options. (iii) Something important is at stake. (iv) You are the only person that has the chance to intervene. These hypothetical scenarios are approached with the very same intuitions we have acquired in and for real-life scenarios. But scenarios like *Footbridge* are different from real-life scenarios in several important ways, and the features that make them different are the same features that render making an intuitive judgment both *necessary* (due to restrictions of time, as specified by feature [i]) and *liable to error* (due to the unusual character of the scenario, as specified by features [ii to iv]). The most plausible interpretation of subjects' performance in Greene's experiments, is not that their deontological judgments are typically unjustified because they are driven by emotions that respond to morally irrelevant factors. Rather, deontological judgments, and moral judgments in general, are based on intuitions which have been developed under real world conditions, and these intuitions are prone to misfire in outlandish moral emergencies like *Footbridge*. This does not, however, entail that deontological judgments always misfire, or even typically do so. Most real-life scenarios are unlike moral emergencies, and our intuitions work quite well in these ordinary cases. It is worth emphasizing that moral intuitions are not special in this respect: we know, for example, that logical intuitions (intuitions about what follows from what) misfire severely and robustly once they are applied to unusual, detached contexts of theoretical reasoning (e.g., the Wason selection task). They work perfectly well, however, in their natural habitat of social reasoning (Cosmides 1989; Cosmides and Tooby 2008).

The reason why Greene's (and, in a similar vein, Singer's) normative argument against the trustworthiness of deontological intuitions fails is not that it rests on the false assumption that the factors that deontological intuitions respond to are morally irrelevant. The claim that whether a morally wrong action is brought about in an up close and personal way or not does not bear on the moral quality of that action is very plausible, after all. It fails because moral judgments do not directly pick up on distinct situational features. What people respond to when assessing real or hypothetical scenarios is the whole narrative of the story that is presented to them. Accordingly, the question whether an intuition is justified or unjustified does not directly depend on whether some of the factors it responds to are morally relevant or not. Rather, it depends on whether that intuition stems from a process of belief formation which is reliable under the circumstances it has been designed to deal with. And since we are talking about *human* moral judgment, this will always involve trade-offs between accuracy and limited information processing abilities.

Intuitions are, unlike some emotions, prime candidates for cognitive penetrability (Gerrans and Kennett 2006). They are typically

- *acquired* rules of thumb, which are
- highly *malleable* and
- amenable to rational *reflection*

that can result in the acquisition of new or, via feedback mechanisms, the improvement of old moral intuitions. Recently, Jillian Craigie has shown that one major flaw in the dual process framework is the tacit assumption that moral judgments are produced by the automatic system *or* the controlled system, as there is "no further discussion of the extent to which moral intuitions are amenable to modification, particularly modification as the result of reflection. The proposal's clear focus on the idea that moral judgments are driven by *either* one system or the other, and the need for reflective processes to override intuitive responses, suggests that the scope for reflective modification of moral intuitions is assumed to be minimal" (Craigie 2011, 60; see also Lapsley and Hill 2008). This interaction between reflective and affective processes, however, is absolutely crucial to distinguish genuine moral judgments from mere preferences or states of (dis) approval.

Metacognition and higher order reflective processes play a central role in moral judgment. Failing to see this is, by the way, no negligible lapse. It is to misunderstand the very practice of moral judgment, which is inextricably tied to the possibility of critically reflecting on one's immediate affectively charged intuitions, to reconsider them if necessary, or to back them up with appropriate further grounding (Jones 2006; Fine 2006; Kennett and Fine 2009; Sauer 2011b). Gerrans and Kennett (2010) argue that Greene's "neurosentimentalism," as they call it, fails to account for this dimension altogether. They hold that genuine moral judgments flow from morally competent subjects. Moral competence, however, cannot be ascribed to agents on the basis of singular instances of putative moral verdicts, but only in a way that takes an agent's past and future into account. Complementarily, genuine moral judgments can only be made by *persons*—not by subpersonal neural systems. Given their amenability to education (past) and reflection (future), judgment formation on the basis of heuristics meets this constraint.

Moral intuitions are System I responses. They are quick, effortless, and cognitively efficient. These features are worth wanting; but they often come with the price of being inflexible. Elisabeth Anderson (2005) has suggested that this need not be so: she argues that moral heuristics should not be understood as rigid rules that mandate a certain response. Rather, they can be seen as inputs into a process of judgment formation, and allow for greater context sensitivity and openness to reflection than do other System I processes.

What is the positive evidence that people's nonconsequentialist judgments are based on educated deontological intuitions? The two most popular explanations of the emergence of deontological moral intuitions appeal to their emotional basis (this is Greene's approach discussed in this chapter) or to an innate set of moral principles—a "universal moral grammar" (Mikhail 2007; Hauser 2006)—to account for the deontological patterns we find in people's moral thinking. I have already discussed many of the issues Greene's dual process theory is confronted with. Nativist proponents of a universal moral grammar, on the other hand, primarily appeal to two arguments to make their case for innate moral rules: first, these rules show up early in children's development—one of the hallmarks of innateness. Second, there are arguments from the so-called *poverty of the (moral) stimulus*. This type of argument holds that the level of complexity and sophistication of many of the moral rules children seem to grasp early on is simply too high for these rules to be extracted from the—admittedly scant—evidence they receive in their environment. Children appreciate the doctrine of double effect before they are given the cues needed to infer the content of this rule.

In a brilliant twist, Shaun Nichols and colleagues (forthcoming) grant these two points, and still manage to explain the emergence of deontological thinking in terms of a rational education of intuitions rather than by recurring to innate modules or brute affective reactions. At the heart of their explanation is a basic principle of statistical learning: the size principle. Here is an intuitive illustration: there is a selection of dice in front of you, one with four sides, another with six, all the way up to ten. One of the dice—you do not know which one—is rolled a few times, and the numbers that come up are all below four. Which die would you guess has been rolled? To be sure, the results are *consistent* with all of them (there are no elevens, for instance); but it would be pure coincidence for the ten-sided die to yield the observed numbers. So you would—and should—guess that the four-sided one was rolled. In short: smaller hypotheses which are consistent with the observed data are, for probabilistic reasons, rationally preferable. It would be *irrational* to avert to a larger hypothesis to explain the observed pattern. In general, this is how children learn words, which is why they only need a few dogs to know what *dog* means without having to rely on any negative examples. They simply jump to the smallest set consistent with the evidence {German Shepherd, Chihuahua, Golden Retriever} given by their teachers.

The important step in their argument is that the size principle can be used to carve up the action space as well. As a matter of principle, the moral domain, and thus the subject matter of moral judgment, could cover all events, including natural ones and the consequences of other people's actions. All events, including natural ones, are the widest possible scope of moral rules. Actions only, including not just mine but other's actions as well, are the next narrowest. Narrowing down this set step by step, there are events caused by one particular agent excluding others, but including accidental outcomes. Merely foreseen outcomes are the next smallest category of events and are both agent caused and nonaccidental. Finally, intended, and not merely foreseen, consequences are a subset of those.

Now, given the size principle as a rule of Bayesian learning, and the evidence children get regarding which subset of all events is governed by moral rules, what would it be rational for children to infer? How wide should they cast their net? In order to answer this question, we need to know what information about the scope of moral rules children *do* get. It turns out, on the basis of evidence from adult-child communications (obtained from the so-called CHILDES database and coded for normative content), that over 99 percent of adult communication about behavioral rules is consistent with a narrow scope interpretation of moral rules according to which they concern *doings*, rather than doings *or* allowings. There are a lot of "Don't do X's" and only very few "Don't let X's happen." From this, children should rationally infer that actions—which are narrower in scope than events that were brought about *or* merely allowed to happen—are where the moral action is.

Finally, which type of moral intuition overlaps with the subset of actions children should rationally infer moral rules to be about? Only deontological rules exclusively target intentional doings, and privilege them over allowings or merely foreseen events, whereas consequentialism sees no difference between, for instance, actions and omissions. This entails that, given the evidence available to children, it would be spectacularly irrational for them to acquire a consequentialist moral framework. The deontological patterns that structure people's moral intuitions are rationally acquired from their infantile learning environment. In my idiom: nonconsequentialist moral judgments are based on rationally educated intuitions.

Conclusion

The content of these intuitions is extracted from real-world situations, which is where children receive their moral education. But consider this potential problem: the educated intuitions account seems inadequate as a model for moral judgments about novel and unrealistic cases that are far removed from real-world situations. A lot of the experimental literature, including the material used to support dual process theory, is about just such cases. The educated intuitions model is thus unlikely to be explanatorily adequate for moral judgments about far-fetched scenarios. But look at it this way: if moral judgments are educated moral intuitions, then this very fact explains people's seemingly poor performance when making moral judgments about novel or outlandish cases, and it does so in terms of the quality of the stimuli used—namely their *external validity* (Bauman et al. 2014)—rather than in terms of allegedly unreliable cognitive processes evidence for the involvement of which is poor anyway.

Conversely, my account suggests that moral judgment should be studied in ecologically more valid contexts. The positive proposal outlined above hinges on the idea that deontological intuitions are generally justified. The fast-and-frugal mechanisms they stem from are the best ones available to us when dealing with real-world issues. In outlandish circumstances, these established methods of judgment formation can be made to severely misfire. One thing the DP model has shown is that people are often fairly bad at adjusting their habitualized responses to new and unusual situations. But one might argue that Greene's model can accommodate all this. It could be that the model holds that deontological intuitions often do have normative force, but that we can single out morally salient cases—the dilemmas Greene and others have studied—in which these intuitions misfire in predictable ways. But this maneuver comes at a very high cost. Restricting the

DP model to the claim that we should be wary of deontological responses in some far-fetched cases renders the model trivial at best: what we end up with is a combination of the claims that, first, deontological intuitions are often based on automatic processes (which does not imply that this is not also the case for consequentialist intuitions), and second, that they misfire in artificial scenarios. What we were promised, however, was that deontological intuitions *as such* are based on dubious affective responses, and that these intuitions *as such* are unjustified. And this promise, it seems, cannot be kept.

In this chapter, I have shown that the argument from morally irrelevant factors is unsound. On a closer look, the empirical evidence in support of its first premise turns out to be weak: the conclusions drawn from neuroimaging rely on a problematic reverse inference, the evidence from brain damage supports the first premise neither empirically nor normatively, and the behavioral evidence from reaction time and cognitive load data collapses as a whole.

The normative argument in support of the second premise also cannot be maintained. It has not been shown that people actually do respond to the factors that are deemed morally irrelevant. The central empirical tenet of the DP model—the claim that deontological intuitions are based on emotional processes—is supposed to indirectly support the normative claim that these processes pick up on morally irrelevant factors, because emotions are thought to be unreliable and potentially distortive ways of arriving at moral judgments. But the fact that deontological judgments can depend on mental processes that operate quickly and automatically does not undermine their normative force, and once the personal/impersonal distinction is dropped and the personal force factor ruled out, there are no candidates left for a debunking explanation of deontological judgments. In fact, the view sketched at the end of this chapter offers a picture of deontological judgment, and moral judgment in general, that manages to explain both why patterns of moral judgment can severely misfire in deviant cases and yet be generally justified.

Conclusion

When Alfred Hitchcock shot his 1940 *Rebecca*, he wanted to stay as faithful as possible to Daphne du Maurier's original story—so much so that he chose to edit the film "in camera," shooting only those scenes which were to appear in the final cut, in the exact order in which they were to do so, all to prevent producer David O. Selznick from making too many changes to Hitchcock's directorial vision. Only one thing was changed: in the film, Rebecca's death is an accident. She and her husband Maxim de Winter get into a fight. She taunts and lies to him about being unfaithful, trying to provoke him to kill her because, unbeknownst to Mr. de Winter, Rebecca has terminal cancer. She falls over backward, hitting her head on a large piece of tackle. In the novel, however, Maxim shoots his wife. But the *Motion Picture Production Code*, (informally known as the *Hays Code* after its chief supervisor William H. Hays) would not allow that a murderer should get away with a crime—and so this part of the plot had to be changed to be fit for the masses, as the people behind the *Code* were afraid that it would offend the audience and ultimately corrupt its sense of right and wrong. Unavenged crimes were not the only thing, by the way, for which the *Hays Code* had taken safety precautions. Neither white slavery nor miscegenation, neither ridicule of the clergy nor drug use, profanity, sexual hygiene and, not least, "excessive and lustful kissing" were considered appropriate (Leff and Simmons 2001, 287). It took only a little less than another thirty years, until the release of Mike Nichols's *Who's Afraid of Virginia Woolf?* in 1966, for the Production Code to become fully obsolete.

As an illustration of how profoundly moral emotions influence our lives, this example might seem petty and insignificant. In many cases, though, moral emotions have a far greater impact on people's lives. This impact goes beyond what we now consider bizarre and ridiculous, and affects things that are far more important than table manners or how many swearwords we get to hear in the latest flick. For many people, morality is a matter of life and death. On May 13, 1988 Stephen Roy Carr—a twenty-eight-year-old loner who lived on South Mountain in Pennsylvania's Michaux State Forest—saw Rebecca Wight and her partner Claudia Brenner, who had had gone on a hiking trip in the Appalachian Trail, having sex in their camp. He happened to carry a .22 caliber rifle, and shot them. Rebecca Wight died of injuries to the head and the liver, Claudia Brenner managed to flee. When Carr was brought to trial and charged with first-degree murder, he asked that his charge be mitigated to voluntary manslaughter because of the overwhelming moral outrage and disgust he experienced upon witnessing the couple engage in lesbian sex (Brenner and Ashley 1995; Pohlman 1999; Nussbaum 2004). The request did not go through, but this does not change the fact that even today, feelings of moral disgust inspire people's actions as well as legislation in many countries. They continue to be defended as valuable moral guides for legal practice (Kass 1997).

Moral emotions are incredibly powerful, for better and worse. At the same time, it is disconcerting just how easily yet profoundly they are affected by the sheer contingency of the circumstances we happen to find ourselves in:

> The point is not that circumstances influence behaviour, or even that seemingly good people sometimes do lousy things. No need to stop the presses for that. Rather, the telling difficulty is just how insubstantial the situational influences effecting troubling moral failures seem to be; it is not that people fall short of ideals of virtue and fortitude, but that they can be readily induced to radically fail such ideals. (Doris and Stich 2005, 119)

One of the most troubling accounts of how this can happen comes from the story of Reserve Police Battalion 101 and Major Wilhelm Trapp—whose popularity with his men had earned him the nickname, "Papa Trapp." In July 1942, Trapp and his 500 men were assigned a special, and indeed especially horrendous task (Browning 1992). They were told to immediately leave their accommodation in the Polish village of Biłgoraj and travel to Józefów. After they had arrived, Major Trapp explained to them that of the 1,800 Jewish inhabitants in the village, the men were to be sent to a concentration camp. The women, children, and those considered unsuitable for hard labor in the camps were to be shot immediately. To those who did not think they were able to carry out the task, Trapp even said they could opt out, an offer that merely a dozen of them accepted (Gigerenzer 2007, 179). The others followed their orders, and went on to fulfill their duty. Their twisted feelings of loyalty had trumped their horror at the assignment. It takes only very little to distort people's common moral sense into an utterly perverted system of morality, and turn a group of ordinary men into mass murderers (Pauer-Studer and Velleman 2010).

But there are also grounds for optimism. Sure, our emotions are easily manipulated. Once we have become aware of this, however, we can use this knowledge to our advantage, and create conditions that will make it less likely for our emotions to lead us astray. It takes only fairly minor changes in people's situational setting for their moral feelings to make them commit horrible deeds. But it takes just as little to lead people and their emotions back on the right track. In fact, when people lose their emotional sensitivity, they also lose their understanding of what the right track is. The developmental psychologist Essi Viding from the University of London works with psychopaths. During one of her interviews, she showed pictures of fearful faces to a psychopathic criminal and asked him whether he could identify the emotion. He said: "I don't know what the emotion is, but it's the face people make right before I stab them" (Viding, pers. comm.; see also Ronson 2011). Robert Hare, one of the leading experts on psychopathy, quotes this astonishing story one psychopath told him about his crimes:

> I was rummaging around when this old geezer comes down the stairs and … uh … he starts yelling and having a fucking fit … so I pop him one in the, uh, head and he still doesn't shut up. I give him a chop to the throat and he … like … staggers back and falls on the floor. He's gurgling and making sounds like a stuck pig (laughs) and he's really getting on my fucking nerves so I … uh … boot him a few times in the head. That shut him up … I'm pretty tired by now, so I grab a few beers from the fridge and turn on the TV and fall asleep. The cops woke me up (laughs). (Hare 1999, 91)

Sometimes, our emotions block our sense of what the right thing to do is; a false sense of tradition and warped feelings of disgust, loyalty, or superiority can make us behave wrongly. But a lack of emotion is not morally desirable, either, and so there is no reason whatever for us to wish that we be freed from our emotional sensitivity altogether.

Moral norms and the feelings that make them resonate in us are so deeply engrained in our everyday actions and experiences that their absence would render successful social interaction impossible. These emotionally supported norms are so obvious to us that we frequently take them for granted. In the 1960s, the sociologist Harold Garfinkel invented the "breaching experiment" in order to make these hidden norms—which he referred to as the moral order as seen "from within" (Garfinkel 1964, 225)—visible and show how social interaction breaks down when people cannot rely on their intuitive sense for a shared normative background anymore. Garfinkel advised his students to enter familiar situations—but act in unfamiliar ways. In one of the assignments, students were requested to behave like strangers in their home. They would be overly polite, distanced, formal, and pretend they did not know anything about their family members' background, habits and preferences. In most cases

> family members were stupefied. ... Reports were filled with accounts of bewilderment, shock, anxiety, embarrassment, and anger and with charges by various family members that the student was mean, inconsiderate, selfish, nasty, or impolite. Family members demanded explanations: What's the matter? What's gotten into you? Did you get fired? Are you sick? ... Are you out of your mind or are you just stupid? (Garfinkel 1964, 232)

In another experiment, students had to start a conversation with someone and, while behaving as if nothing unusual was happening, move their face toward the other person until their noses were almost touching. Needless to say that people, especially men, did not like this at all, and in many cases it had become impossible to restore normal interaction afterwards, even when it had been explained that the incident had been part of a sociological experiment. What all these experiments have in common is how amazingly cringe-worthy they are, how awkward they make us feel even upon reading about them, how outraged the involuntary participants of these experiments became when someone seemed to have lost his understanding of the moral fabric of society, and how much strength it cost those scientists to overcome their own emotions and break the very norms whose authority they were brought up to take for granted.

These are just some examples of how moral emotions influence the everyday lives of individual people. But they also affect society as a whole: what emotions one is susceptible to influences the profession one chooses and the people one meets. Emotions have a massive impact on the political outlook one is likely to acquire and on who we trust and vote for. What seemed like the most impractical concern can suddenly become very practical indeed: Max Weber's (1988) narrative, for example, has it that your eschatological views affect how well you do in business. In the long run—and we all bear witness to the reality of this scenario—this can change the fate of whole civilizations. Social psychologists have found that what you care most about, what you find offensive, corrupt, or simply disgusting best predicts what political views you have. Sensitivity to disgust accurately predicts whether people disapprove of homosexuality (Inbar, Pizarro, and Bloom 2009), and there is a strong general correlation between where someone is located on the political spectrum and the rigidity of someone's views about sexual morality (Haidt and Hersh 2001). The most central division dominating politics in the West—the one between the left and the right, or liberals and conservatives—in the end seems to boil down to differences in people's tendency to moralize about certain issues or not. Liberals and people on the left of the political spectrum think about morality in terms of harm and fairness. They do not, as conservatives and people from the political right, consider respect for authority, loyalty to one's particular community, or a sense for the purity and sacredness of the human body to be issues of genuine moral significance (Haidt 2007; Graham, Haidt, and Nosek 2009). Liberals just do not feel strongly about these things. In fact, many of them consider experimenting with food or sex, questioning the authority of traditions or transcending the cultural context one was born into to be valuable pursuits. However, although these observations are now turning more and more into accepted wisdom in social psychology, they are only one side of the story. It would be a gross exaggeration to say that people who identify as liberals lack a sense of community, authority, or purity. They merely apply these senses to different types of things than conservatives. Liberals can sometimes be snobs, and they are happy to moralize about people who watch soap operas and eat unhealthy or processed food. They are disgusted and outraged by hate crimes, they respect the voice of those moral authorities that fight for the poor and downtrodden, and they share a sense of community and feel solidarity with the victims of discrimination, oppression, and exploitation, regardless of where they come from and what group they belong to. It remains true that many political differences reflect emotional differences. But these differences are less due to what moral emotions people are capable of, and more due to what issues and social spheres they apply them to.

To be sure, there are limitations to whom one empathizes with, and how strongly one can be motivated to care about the well-being of others. If I told you that yesterday, I drove by a shallow pond and saw a child that was in danger of drowning but did not save her because I did not want to muddy my expensive suit, you would think I am a monster. If I told you that yesterday, I heard about the dire situation of starving children in a foreign country, but did not donate any money because I was planning to buy an iPad that same day, you'd probably cut me some slack. And yet the two cases do not seem so different, except for the fact that the first child was close to me and the others were far away (Singer 1972, 2005). Our moral and political views about topics such as developmental aid are not just influenced by how internal personal dispositions shape our emotions. They are often deeply affected by external situational factors such as familiarity or spatial distance. When people have difficulties empathizing with others, they can omit to do the right thing and not give to famine relief. But they can also actively start doing the wrong thing. We can be stuck in inaction because the hardship of others is too far away from us, or it is suffered by people we perceive merely as representatives of a group rather than individuals (Unger 1996). In other cases, we can become capable of actively hurting others because we have come to see them as subhuman vermin, or because large bureaucratic organizations dilute personal responsibility and sophisticated technologies turn heinous decisions into apparently simple technical options (Bauman 1989; Foucault 1977). What we should learn from these observations in particular and the empirical psychology of moral judgment in general is not so much what the psychological basis of moral judgments is, but that we ought to be very careful when making moral judgments. The available evidence can assist us in being careful, because it gives us knowledge about which irrelevant things our judgments are most likely to be influenced by.

I began the sixth chapter of this book with a discussion of disgust and ended it with an assessment of the importance of reflection. In making moral judgments, these are two sides of the extreme we constantly have to navigate between: disgust is a visceral response that is part of our animal nature; reflection is a calm intellectual endeavor that, some believe, removes us from and elevates us above our animal nature. This dialectic of the visceral and the cerebral is the common thread in all preceding chapters, and the main question that concerned me in this book was: which should we trust—the gut or the head? Which is prior—intuition or reason?

Take disgust again. Some (Rozin and Fallon 1987; Kelly 2011) argue that we should be especially skeptical of the moral significance of disgust. Perhaps this feeling is untrustworthy because of its origins as a food rejection device for omnivores, which then "spilled over" into the sociomoral domain. Martha Nussbaum has suggested that things such as feces, blood, and semen remind us of our animal nature, and therefore disgust us. But this feeling can grow out of proportion, and can make us develop an irrational loathing toward our own humanity.

Others have argued that this could not be further from the truth, and that when it comes to our most central moral convictions, it is reason and reflection we should often be suspicious of. Leon Kass even goes so far to suggest that there is something problematic about the very attempt to argue rationally about some moral judgments, as if this were a sign of a corrupt mind: "we are suspicious of those who think that they can rationalize away our horror, say, by trying to explain the enormity of incest with arguments only about the genetic risks of inbreeding" (Kass 1997, 20). Similar disagreements arise about other emotions and the degree to which they should be moderated and filtered by reflection. Some praise empathy and compassion as the foundation of morality; others reject it as the source of partiality and parochialism. Some welcome a feeling of guilt as an inner moral compass; some hold it in contempt for the suffering it can cause. Finally, some applaud our ability to act on the basis of principles, even against the voice of our feelings; some look at this ability with horror, and blame it for many of the gravest atrocities ever committed in the name of morality.

In many ways, the position I have defended offers something of a compromise, and aims to split the difference between the friends and foes of intuition. I have argued that moral judgments are made automatically, and that we do and ought to rely on our automatic intuitions in moral matters. However, I insisted that this is not because there is a special kind of wisdom embodied in our gut reactions, but rather because these moral intuitions did not remain unaffected by reasoning, which figures in their acquisition, formation, maintenance, and reflective correction. On the other hand, I have argued that reasoning and reflection are indispensable for moral judgment. However, I emphasized that this is not because reason has any special moral authority, but rather because it is a tool us humans have been uniquely equipped with that allows us to improve upon our habitualized patterns of moral judgment.

My main claim has been that the practice of moral judgment deserves to be described as an exercise of reason. Throughout this book I have intended this to be a metaethical claim. But now, at the end of this book, I am inclined to betray the promise I made early on in the introduction, and to let the normative cat—*my* normative cat—out of the metaethical bag: I am convinced that the interaction of emotion, intuition, and reason makes moral progress not just possible, but actual. That many of our feelings of revulsion have been demoralized is progress; we are making progress by moving further and further away from the parochialism of our sympathies; and it is progress that we have come to realize that which tool one uses to eat is really not worth the agitation. These different threads form part of one big tapestry of progress. Some refer to it as a progress of civilization; others call it an advance in our moral knowledge. Still others prefer metaphors and describe it as enlightenment. I stick to tradition and call it a progress of *reason*.

Notes

Introduction

1. Some might think that table manners are not really *moral* rules at all. I briefly return to this entirely legitimate worry at the end of this introduction. The above account of the princess and the fork can be found in Elias 1994.

2. This does not mean, however, that I have nothing to say about the theories developed by Greene, Nichols, or Hauser and Mikhail. In fact, the first chapter of this book deals extensively with Greene's dual process model of moral cognition; the second chapter explains what the theory of Universal Moral Grammar has to say about the importance of conscious, principle-based reasoning for moral judgment; and finally, in chapter 7, I will draw on Nichols's notion of "core moral judgment" to explain why emotions may be necessary for moral judgment.

1 The Effectiveness of Moral Reasoning

1. In a recent paper, Schwitzgebel and Ellis (forthcoming) operate with a similar account of the difference between genuine reasoning and rationalization.

2 Educated Intuitions

1. My argument will not in any way rely on the claim that philosophers of action and empirical moral psychologists *explicitly endorse* the incompatibility thesis (although some might). Rather, it seems to be the case that many researchers are often implicitly, and sometimes maybe even inadvertently, *committed* to the thesis, because most available theories of rational action and moral judgment are, as a matter of fact, unable to account for the possibility of automatic yet rational thought and behavior.

2. It is somewhat doubtful whether there are any rationalists out there who would recognize their position in Haidt's description of rationalism. I do not think, however, that Haidt is arguing against a straw man. There is at least some plausibility in

the idea that we think before we judge morally, and there is an undeniable tendency among people to overestimate the extent to which that happens.

3. This will be explained in more detail in chapter 7.

4. In his more recent writings, Velleman (2010) seems to be more aware of this.

5. In recent years, this idea—that a great many automatic processes, especially habits, can qualify as rational—has become increasingly popular. John McDowell has championed the idea. He argues that habits are the key to bridging the gap between the space of causes and the space of reasons, and has defended the view that conceptual capacities—whose involvement seems to be necessary to anything that can be subject to rational justification—can become embodied in the "habits of thought and action" which, over time, become "second nature" (McDowell 1994, 84). Following McDowell, Sabina Lovibond argues that the capacity to make moral judgments and to act on them essentially consists in the initiation into a culture, and the acquisition of habits and traits that come with the participation in social practices:

> Over time, our participation in these activities ... gives rise to a "second," or acquired, nature. This second nature is manifested in behaviour which, though learned, is largely unreflective ...; and which if we do make it into an object of reflection, usually produces in us a sense of inevitability. From one point of view, the dispositions that constitute our second nature are passive, for they are dispositions to be affected in a certain way: ideally, to register the "proper force and necessity" of reasons for judgment. ... However, it is a feature of human socialization ... that one is led not just to receive and process sensory input from one's environment, but to recognize the state of the world as imposing rational constraints one one's thinking. (Lovibond 2002, 25–26)

One of the main goals of moral upbringing is to equip subjects with the capacity to make good moral judgments. And if moral judgments, as the empirical evidence suggests, largely depend on intuitive processes, then an understanding of how the education of moral intuitions works is of foremost importance.

6. Snow explains this concept as follows: "A chronically held goal is enduring or long-lived. A goal the mental representation of which is often activated by the appropriate stimuli becomes chronically accessible in the sense that it becomes readily activatable. In goal-dependent automaticity, goal activation occurs outside of the person's conscious awareness through encounters with triggering stimuli" (2006, 547).

7. An interesting philosophical discussion of cases where emotions (rather than explicitly articulated moral principles) do a better job at latching onto the features of the world that constitute (moral) reasons for acting can be found in Arpaly 2003.

3 Moral Reasoning as a Social Practice

1. Similar accounts can be found in Peirce (1877), Heidegger (1927), Wittgenstein (1969), Austin (1979), Strawson (1962), Wallace (1994), Willaschek (2002, 2007), and Thomas (2006, 2010). The term "challenge and response" was coined by Wellmann (1971).

2. This example refers to Haidt, Björklund, and Murphy's (2000) unpublished study, in which they asked experimental subjects to sign a contract stating that they would sell their soul for two dollars.

3. Huemer (2005) suggests that the practice of judgment in general presupposes something like his "principle of phenomenal conservativism."

4. This distinction figures prominently in the work of Strawson (1962), Austin (1979), Wallace (1994).

5. A defeater is constituted by "information that can mandate the retraction of the conclusion of the defeasible argument" (Pollock 1986, 4). P is a *rebutting* challenge for Q iff $(P \rightarrow \neg Q)$; P is an *undercutting* challenge for Q iff $(R \rightarrow Q) \wedge [P \rightarrow \neg(R \rightarrow Q) \vee P \rightarrow \neg R]$.

6. Skeptical conclusions readily follow once one has discovered that many ordinary beliefs cannot be based on those purportedly more basic beliefs. This line of reasoning is diagnosed as "epistemological realism" and criticized by Michael Williams (1988).

4 Moral Error

1. This and other sacrificial dilemmas of the "trolleyological" are discussed in more detail in the final chapter.

2. The phenomenon of psychopathy and the implications it has for metaethics and moral psychology will be discussed in chapter 6.

3. Egan (2007) criticizes expressivist theories of moral judgment on grounds which are similar to mine. His argument, however, is based on semantic considerations, whereas mine starts from more broadly psychological ones. And although Egan and I agree that expressivist and sentimentalist theories of moral judgment, respectively, have to struggle with infallibility, there are important differences when it comes to where this infallibility stems from. In the case of my argument, as we will see below, it stems from the close connection sentimentalists establish between subjects' moral judgments and the response-dependent moral properties their judgments literally bring into existence. Egan argues that the way quasi-realist expressivists have devised for dealing with attributions of moral error to other subjects or one's own past or future selves does not work equally well for cases of self-concerned epistemic

modesty, in which a subject concedes that a moral judgment of her present self might be wrong. If A attributes to B the moral judgment that *p*, while expressing herself that not p, then the quasi realist's analysis of this judgment amounts to: "B approves of *p*, and I (=A) disapprove of *p*." However, in the case of one's *present* self, this type of analysis yields an unacceptable result, because it turns worries about whether one is right into worries about whether one knows what one presently (dis) approves of. This charge makes two claims at once which I prefer to treat separately: first, it mirrors what I call the infallibility problem, and second, it resembles my observation that introspective mistakes about what emotional dispositions one has or not are the "wrong" kind of mistake.

4. For the sake of brevity, I use the term "constitution" in this working definition of emotionism. In chapters 7 and 8, I will explain why the best reading of this constitution relation is that emotional responses are both necessary and sufficient for moral judgment. The details of this do not yet matter here, however, as I am not concerned with the problems PT or MT might have *individually*, but merely with the problems that arise from their *relation* to each other.

5. In chapters 7 and 8, I explain in more detail why a rationalist account of moral judgment need not explain away the significance of emotions for moral judgment.

6. This last point could be thought to hold for constructivist theories as well. These theories, however, do not aim to offer an account of the psychology of moral judgments or the metaphysics of value, and thus do not have to struggle with the problem of moral error in the same way. Take Rawls's theory of justice: this theory does not hold that subjects make moral judgments by employing the veil of ignorance procedure. Nor does it claim that moral facts (whose existence constructivists deny) are constituted by the procedure.

7. The examples I give here are not, nor need they be, uncontroversial. For those who disagree with any of (1)–(6), the negation of the respective example will work equally well.

5 Are Emotions Necessary for Moral Judgment?

1. Phineas Gage was a mid-nineteenth-century construction worker for a New England railroad company who suffered a massive injury to his frontal lobe. While working on building a railroad, he inadvertently caused an explosion which send a large iron rod through his forehead. His case gained prominence in popular as well as academic circles not only because Gage survived this incident, but also because of the change in his character that occurred to him despite the fact that he retained full functionality in virtually all other cognitive respects. The most famous account of this case can be found in Damasio 1994.

2. I am sidestepping many conceptual issues with defining what an emotion is, and with distinguishing it from feelings, moods, or affects. The main reason for this is that it is very hard to come up with such a definition that remains noncommittal with respect to the various available theories of the emotions. In what follows, I will therefore mostly work with particular examples of emotions, and hope to be able to proceed in this bottom-up fashion.

3. This last point might seem like too strong an assumption to make, and perhaps it is. What it can serve to illustrate, however, are the limitations of the empirical evidence in cases such as these. It is worth emphasizing that, if primates really did have the necessary prerequisites for full-blown moral judgment, then they would presumably make full-blown moral judgments. But they don't.

6 Are Emotions Sufficient for Moral Judgment?

1. Eskine and his colleagues implicitly grant this point when they say that gustatory disgust constitutes "extraneous sensoriperceptual information during moral processing" (2011, 297).

2. Speaking of moral judgments in terms of "thoughts" seems wrong in this context, because emotionists reject the idea that moral judgments have a cognitive foundation. My use of the word "thought" is more akin to what Sneddon refers to as "verbalizations" of internal emotional dispositions (2011, 31).

3. Roughly, what I have in mind here is that a method of judgment-formation is sensitive if it satisfies the following "Nozickian" condition: a judgment that p is sensitive iff $\sim p \rightarrow \sim B_x(p)$ and $p \rightarrow B_x(p)$ (cf. Nozick 1981, 176). Both conditions are important because, given the empirical evidence, disgust responses that pick up on extraneous features are ruled out by the first conjunct, cases of artificial mood induction that prevents people from picking up on morally relevant features are ruled out by the second conjunct. For an overview of accounts such as these, see Pritchard 2005.

4. Recently, John Doris (2015) has proposed a similar account, tying the exercise of agency to whether a subject's actions have been caused by processes which overlap with the justifying reasons she would cite in light of possible defeaters.

7 Educated Deontological Intuitions

1. Utilitarianism is of course only one possible version of consequentialism. But since the details of this do not matter for the argument of this chapter, and since Greene himself does not seem to give much weight to this distinction, I will use utilitarianism and consequentialism interchangeably.

2. To my knowledge, the first use of trolley cases in the experimental literature can be found in Petrinovich, O'Neill, and Jorgensen 1993. I apologize for not being able to remember whom to thank for this pointer.

3. Here they are, in the version Greene uses in his experiments:

Trolley

You are at the wheel of a runaway trolley quickly approaching a fork in the tracks. On the tracks extending to the left is a group of five railway workmen. On the tracks extending to the right is a single railway workman. If you do nothing the trolley will proceed to the left, causing the deaths of the five workmen. The only way to avoid the deaths of these workmen is to hit a switch on your dashboard that will cause the trolley to proceed to the right, causing the death of the single workman. Is it appropriate for you to hit the switch in order to avoid the deaths of the five workmen?

Footbridge

A runaway trolley is heading down the tracks toward five workmen who will be killed if the trolley proceeds on its present course. You are on a footbridge over the tracks, in between the approaching trolley and the five workmen. Next to you on this footbridge is a stranger who happens to be very large. The only way to save the lives of the five workmen is to push this stranger off the bridge and onto the tracks below where his large body will stop the trolley. The stranger will die if you do this, but the five workmen will be saved. Is it appropriate for you to push the stranger on to the tracks in order to save the five workmen?

(These dilemmas can be found in the supplementary material to Greene et al. 2001; see 2107n8.)

4. The load condition consisted in a task to detect the cipher "5" in a stream of numbers scrolling across the screen on which the dilemmas were presented.

5. The results are not unambiguous, however. Eight out of the twenty-one moral scenarios yielded no difference in responses between VMPFC patients and two control groups of normal, healthy individuals and people with other types of brain damage. For the remaining thirteen scenarios, significant differences were found. Although all subjects exhibited similar tendencies with respect to each of the scenarios, patients with damage to the VMPFC were indeed more likely to judge it appropriate to perform an emotionally conflicting action (for example, to smother one's baby to protect a group of people from a lethal threat).

6. Here is this particularly ominous example:

Hired Rapist

You have been dissatisfied with your marriage for several years. It is your distinct impression that your wife no longer appreciates you. You remember how she appreciated you years ago when you took care of her after she was mugged. You devise the following plan to regain your wife's affection. You will hire a man to break into your house while you are away. This man will tie up your wife and rape her. You, upon hearing the horrible news, will return swiftly to her side, to take care of her and comfort her, and she will once again appreciate you. Is it appropriate for you to hire a man to rape your wife so she will appreciate you as you comfort her?

(This dilemma can be found in the supplementary material to Greene et al. 2001; see 2107n8.)

7. To show very clearly what has been found, I shall sum up the results in headline style:

(i) Cognitive load does *not increase the frequency* deontological judgments. In fact, it does slightly increase the frequency of utilitarian judgment (60 percent vs. 61 percent).
(ii) In the absence of load *there is no RT difference* between deontological and utilitarian judgments. (This confirms McGuire's and his colleagues' analysis from above.)
(iii) Cognitive load does *slow down* utilitarian judgment. (It should be noted, however, that the effect is not spectacular: 6.5 seconds for utilitarian vs. 5.8 seconds for deontological judgments on average.)

To make sense of these puzzling results, Greene reanalyzed the data after dividing his eighty-two participants into two groups, based on their general tendency to prefer utilitarian or deontological judgments. For these "high" and "low utilitarian" groups, the results were:

(iv) Cognitive load *does not have a significant effect* on which judgments people make. (Low utilitarian group: 43 percent utilitarian judgment under load, 41 percent in the absence of load; High utilitarian group: 79 percent utilitarian judgment under load, 78 percent in the absence of load.)
(v) Cognitive load *slows down utilitarian judgment* in both groups.

Furthermore, the RT effect originally predicted was eventually found in low utilitarian subjects, but the opposite held true of high-utilitarian participants:

(vi) In the absence of load, *low utilitarian subjects take longer* to make utilitarian judgments.
(vii) In the absence of load, *high utilitarian subjects* make utilitarian judgments *more quickly*.

8. Here is one of their stories, which I shall refer to as *Heavy Object*:

Heavy Object

Ned is walking near the train tracks when he notices a train approaching out of control. Up ahead on the track are 5 people. Ned is standing next to a switch, which he can throw to turn the train to an aside track. There is a heavy object on the side track. If the train hits the object, the object will slow the train down, giving the men time to escape. The heavy object is 1 man, standing on the side track. Ned can throw the switch, preventing the train from killing the 5 people, but killing the 1 man. Or he can refrain from doing this, letting the 5 die. Is it morally permissible for Ned to throw the switch? (Hauser et al. 2007, 6)

9. More precisely, Greene has replaced his old "Me Hurt You" criterion with the PF factor, which are two ways to *spell out* the content of the personal/impersonal distinction. For reasons of simplicity, I shall refer to the original proposal simply as the personal/impersonal distinction.

10. *Footbridge Pole*

An empty runaway trolley is speeding down a set of tracks toward five railway workmen. There is a footbridge above the tracks in between the runaway trolley and the five workmen. On this footbridge is a railway workman wearing a large, heavy backpack. If nothing is done, the trolley will proceed down the main tracks and cause the deaths of the five workmen. It is possible to avoid these five deaths. Joe is a bystander who understands what is going on and who happens to be standing right behind the workman on the footbridge. Joe is near a six-foot-long pole. Joe sees that he can avoid the deaths of the five workmen by using the pole to push the workman with the heavy backpack off of the footbridge and onto the tracks below. The trolley will collide with the workman, and the combined weight of the workman and the backpack will be enough to stop the trolley, avoiding the deaths of the five workmen. But the collision will cause the death of the workman with the backpack. ... Is it morally acceptable for Joe to use the pole to push the workman off of the footbridge in order to avoid the deaths of the five workmen, causing the death of the single workman instead?

(This dilemma can be found in the supplementary material to Greene et al. 2009.)

11. *Footbridge Switch* and *Remote Footbridge*, the other two dilemmas that were used in this study, are variations of the trapdoor case which differ with respect to how close the person flipping the switch stands to the person dropped onto the tracks.

12. These cases recreate the *Trolley* and the *Footbridge* dilemma using teacups. A mother briefly leaves her two children unattended and specifically tells them not to break any teacups. After he returns from getting a snack, Billy, the older brother, finds that his younger sister Ann has put five teacups on the tracks of his train set. In one case, Billy can divert the train onto a side track with only one teacup on it. In another case, the older sibling (who is now called Susie) can prevent the train from breaking the five cups by throwing another cup at the train. (See Nichols and Mallon 2006)

13. Guy Kahane and his colleagues (2012) also rely heavily on neuroimaging data, the relevance of which seems doubtful to me. For this reason, I will not discuss it any further here. See sections 3 and 5 above.

14. Interestingly, this point is already implicit in Greene's 2001 study, whose findings could only be obtained by including the aforementioned immoral vignettes such as the *Hired Rapist* and others. See section 4 above.

15. Several other attempts have been made at justifying the deontological option in the footbridge case. Robert Nozick (1993, 60) has argued that deontological intuitions stem from human risk aversion. In assessing the dilemmas, people choose different baselines as their starting points, and judge possible gains and losses against the background of those baselines. One possibly lost life against the baseline of five certainly lost lives (in *Trolley*) thus seems more attractive than the possibility of five gained lives against the baseline of one certain death (in *Footbridge*). But Nozick's account doesn't really show that deontological judgments are *justified*. For the most part, he offers an explanation for why people judge the way they do that refers to the influence of framing effects on their judgments.

Frances Kamm (2007) has argued that the nonconsequentialist response is justified in light of the principle of permissible harm. The problem with her account is it is hard to see how to justify her proposal other than through an appeal to the intuition it is supposed to justify: the one that is typically elicited by *Footbridge* and variations of it.

16. A large portion of this chapter is dedicated to an argument against the claim that deontological intuitions are emotional, and thus seems to commit the same kind of unmotivated conflation. That part of the argument is meant for those who think that there is something intrinsically dubious about emotions as such. I do not think that this is the case, but have aimed to show that even if one shares this assumption, one need not agree that deontological judgments have no normative force.

References

Aguirre, G. K. (2003). Functional imaging in behavioral neurology and neuropsychology. In T. E. Feinberg & M. J. Farah (Eds.), *Behavioral neurology and neuropsychology* (pp. 85–97). New York: McGraw-Hill.

Aharoni, E., Sinnott-Armstrong, W., & Kiehl, K. A. (2012). Can psychopathic offenders discern moral wrongs? A new look at the moral/conventional distinction. *Journal of Abnormal Psychology, 121*(2), 484–497. doi:10.1037/a0024796.

Alfano, M. (2009). A danger of definition: Polar predicates in metaethics. *Journal of Ethics and Social Philosophy, 3*(3), 1–13.

Alfano, M. (2011). Expanding the situationist challenge to responsibilist virtue epistemology. *Philosophical Quarterly, 62*(247), 223–249. doi:10.1111/j.1467-9213.2011.00016.x.

Alfano, M. (2013). *Character as moral fiction*. Cambridge: Cambridge University Press.

Anderson, E. (2005). Moral heuristics: Rigid rules or flexible inputs in moral deliberation? *Behavioral and Brain Sciences, 28*(4), 544–545. doi:10.1017/S0140525X05240098.

Anderson, E. (2010). *The imperative of integration*. Princeton, NJ: Princeton University Press.

Anderson, E. (2014). Social movements, experiments in living, and moral progress: Case studies from Britain's abolition of slavery. The Lindley Lecture, February 11, 2014. The University of Kansas.

Andreou, C. (2007). Morality and psychology. *Philosophy Compass, 2*(1), 46–55. doi:10.1111/j.1747-9991.2006.00055.x.

Annis, D. B. (1978). A contextualist theory of epistemic justification. *American Philosophical Quarterly, 15*(3), 213–219.

Appiah, K. A. (2008). *Experiments in ethics*. Cambridge, MA: Harvard University Press.

Arpaly, N. (2000). On acting rationally against one's best judgment. *Ethics, 110*(3), 488–513. doi:10.1086/233321.

Arpaly, N. (2003). *Unprincipled virtue: An inquiry into moral agency*. New York: Oxford University Press.

Arpaly, N., & Schroeder, T. (2012). Deliberation and acting for reasons. *Philosophical Review, 121*(2), 209–239. doi:10.1215/00318108-1539089.

Asch, S. (1955). Opinions and social pressure. *Scientific American, 193*(5), 31–35. doi:10.1038/scientificamerican1155-31.

Asch, S. (1956). Studies of independence and conformity: A minority of one against a unanimous majority. *Psychological Monographs, 70*(9), 1–70. doi:10.1037/h0093718.

Audi, R. (2005). *The good in the right: A theory of intuition and intrinsic value*. Princeton, NJ: Princeton University Press.

Austin, J. L. (1979). A plea for excuses. In J. O. Urmson & G. J. Warnock (Eds.), *Philosophical papers* (pp. 175–204). New York: Oxford University Press.

Ayer, A. J. (1952). *Language, truth, and logic*. New York: Dover.

Bandura, A., Barbaranelli, C., Caprara, G. V., & Pastorelli, C. (1996). Mechanisms of moral disengagement in the exercise of moral agency. *Journal of Personality and Social Psychology, 71*(2), 364–374. doi:10.1037/0022-3514.71.2.364.

Bauman, C. W., McGraw, P., Bartel, D., & Warren, C. (2014). Revisiting external validity: Concerns about trolley problems and other sacrificial dilemmas in moral psychology. *Social and Personality Psychology Compass, 8*(9), 536–554. doi:10.1111/spc3.12131.

Bargh, J. A. (1994). The four horsemen of automaticity. In R. S. Wyer & T. K. Srull (Eds.), *Handbook of social cognition* (pp. 1–40). Hillsdale, NJ: Erlbaum.

Bargh, J. A., & Chartrand, T. L. (1999). The unbearable automaticity of being. *American Psychologist, 54*(7), 462–479. doi:10.1037//0003-066X.54.7.462.

Baron, J. (1998). *Judgment misguided: Intuition and error in public decision making*. New York: Oxford University Press.

Baron-Cohen, S. (2009). Autism: The empathizing-systemizing (E-S) theory. *Year in Cognitive Neuroscience, 1156*, 68–80. doi:10.1111/j.1749-6632.2009.04467.x.

Bartels, D. M. (2008). Principled moral sentiment and the flexibility of moral judgment and decision making. *Cognition, 108*(2), 381–417. doi:10.1016/j.cognition.2008.03.001.

Bartels, D. M., & Pizarro, D. A. (2011). The mismeasure of morals: Antisocial personality traits predict utilitarian responses to moral dilemmas. *Cognition, 121*(1), 154–161. doi:10.1016/j.cognition.2011.05.010.

Bauman, Z. (1989). *Modernity and the Holocaust*. Cambridge: Polity Press.

Bellah, R., Madsen, R., Sullivan, W. M., Swidler, A., & Tipton, S. M. (1985). *Habits of the heart: Individualism and commitment in American life*. Berkeley, CA: University of California Press.

Benbaji, H. (2013). How is recalcitrant emotion possible? *Australasian Journal of Philosophy, 91*(3), 577–599. doi:10.1080/00048402.2012.699078.

Berker, S. (2009). The normative insignificance of neuroscience. *Philosophy & Public Affairs, 37*(4), 293–329. doi:10.1111/j.1088-4963.2009.01164.x.

Besser-Jones, L. (2012). The role of practical reason in empirically informed moral theory. *Ethical Theory and Moral Practice, 15*(2), 203–220. doi:10.1007/s10677-011-9284-9.

Białek, M., & De Neys, W. (2016). Conflict detection during moral decision-making: Evidence for deontic reasoners' utilitarian sensitivity. *Journal of Cognitive Psychology*. doi:10.1080/20445911.2016.1156118.

Blackburn, S. (1993). *Essays in quasi-realism*. New York: Oxford University Press.

Blackburn, S. (1998). *Ruling passions: A theory of practical reasoning*. New York: Oxford University Press.

Blair, J., Mitchell, D., & Blair, K. (2005). *The psychopath: Emotion and the brain*. Oxford: Blackwell.

Blair, R. J. R. (1995). A cognitive developmental approach to morality: Investigating the psychopath. *Cognition, 57(1)*, 1–29. doi:10.1016/0010-0277(95)00676-P.

Blake, P. R., McAuliffe, K., Corbit, J. Callaghan, T. C., Barry, O. ... Warneken, F. (2015). The ontogeny of fairness in seven societies. *Nature, 528(7581)*, 258–261. doi:10.1038/nature15703.

Bonjour, L. (1976). The coherence theory of empirical knowledge. *Philosophical Studies, 30*(5), 281–312. doi:10.1007/BF00357928.

Bortolotti, L. (2011). Does reflection lead to wise choices? *Philosophical Explorations, 14*(3), 297–313. doi:10.1080/13869795.2011.594962.

Brady, M. S. (2008). The irrationality of recalcitrant emotions. *Philosophical Studies, 145*(3), 413–430. doi:10.1007/s11098-008-9241-1.

Brandom, R. (1994). *Making it explicit: Reasoning, representing, and discursive commitment*. Cambridge, MA: Harvard University Press.

Brandt, R. B. (1979). *A theory of the good and the right*. Oxford: Oxford University Press.

Brenner, C., & Ashley, H. (1995). *Eight bullets: One woman's story of surviving anti-gay violence*. Ann Arbor, MI: Firebrand.

Broome, J. (2004). Reasons. In J. Wallace, M. Smith, S. Scheffler, & P. Pettit (Eds.), *Reason and value: Themes from the moral philosophy of Joseph Raz* (pp. 28–55). Oxford: Oxford University Press.

Brosnan, S. J. (2006). Nonhuman species' reactions to inequity and their implications for fairness. *Social Justice Research, 19*(2), 153–185. doi:10.1007/PL00022136.

Brosnan, S. J., & de Waal, F. (2003). Monkeys reject unequal pay. *Nature, 425(6955*), 297–299. doi:10.1038/nature01963.

Browning, C. (1992). *Ordinary men: Reserve Police Battalion 101 and the final solution in Poland*. New York: HarperCollins.

Campbell, R., & Kumar, V. (2012). Moral reasoning on the ground. *Ethics, 122*(2), 273–312. doi:10.1086/663980.

Cassam, Q. (2007). *The possibility of knowledge*. Oxford: Oxford University Press.

Chapman, H. A., Kim, D. A., Susskind, J. M., & Anderson, A. K. (2009). In bad taste: Evidence for the oral origins of moral disgust. *Science, 323(5918*), 1222–1226. doi:10.1126/science.1165565.

Chisholm, R. (1977). *Theory of knowledge*. Englewood Cliffs: Prentice-Hall.

Chomsky, N. (2009). *Cartesian linguistics: A chapter in the history of rationalist thought*. Cambridge: Cambridge University Press.

Ciaramelli, E., Muccioli, M., Làdavas, E., & di Pellegrino, G. (2007). Selective deficit in personal moral judgment following damage to ventromedial prefrontal cortex. *Social Cognitive and Affective Neuroscience, 2*(2), 84–92. doi:10.1093/scan/nsm001.

Cima, M., Tonnaer, F., & Hauser, M. D. (2010). Psychopaths know right from wrong but don't care. *Social Cognitive and Affective Neuroscience, 5*(1), 59–67. doi:10.1093/scan/nsp051.

Clark, A. (2010). *Memento*'s revenge: The extended mind, extended. In R. Menary (Ed.), *The extended mind* (pp. 43–67). Cambridge, MA: MIT Press.

Clarke, S. (2008). SIM and the city: Rationalism in psychology and philosophy and Haidt's account of moral judgment. *Philosophical Psychology, 21*(6), 799–820. doi:10.1080/09515080802513250.

Cohen, S. (1986). Knowledge and context. *Journal of Philosophy, 83(10*), 574–583. doi:10.2307/2026434.

Cosmides, L. (1989). The logic of social exchange: Has natural selection shaped how humans reason? Studies with the Wason selection task. *Cognition, 31(3*), 187–276. doi:10.1016/0010-0277(89)90023-1.

Cosmides, L., & Tooby, J. (2008). Can a general deontic logic capture the facts of human moral reasoning? How the mind interprets social exchange rules and detects cheaters. In W. Sinnott-Armstrong (Ed.), *Moral psychology*, Vol. 1: *The evolution of morality: Adaptations and innateness* (pp. 53–119). Cambridge, MA: MIT Press.

Craigie, J. (2011). Thinking and feeling: Moral deliberation in a dual-process framework. *Philosophical Psychology, 24*(1), 53–71. doi:10.1080/09515089.2010.533262.

Crisp, R. (2000). Review of "Value ... and what follows. By Joel Kupperman." *Philosophy, 75*, 458–462.

Crockett, M., Clark, L., Hauser, M. D., & Robbins, T. W. (2010). Serotonin selectively influences moral judgment and behavior through effects on harm aversion. *Psychological and Cognitive Sciences, 107*(40), 17433–17438. doi:10.1073/pnas.1009396107.

Cushman, F., Young, L., & Hauser, M. D. (2006). The role of conscious reasoning and intuition in moral judgment: Testing three principles of harm. *Psychological Science, 17*(12), 1082–1089. doi:10.1111/j.1467-9280.2006.01834.x.

Damasio, A. (1994). *Descartes' error: Emotion, reason, and the human brain*. London: Penguin Books.

Damasio, A., Tranel, D., & Damasio, H. (1990). Individuals with sociopathic behavior caused by frontal damage fail to respond autonomically to social stimuli. *Behavioural Brain Research, 41*(2), 81–94. doi:10.1016/0166-4328(90)90144-4.

Damm, L. (2010). Emotions and moral agency. *Philosophical Explorations, 13*(3), 275–292. doi:10.1080/13869795.2010.501898.

Dancy, J. (2000). *Practical reality*. Oxford: Oxford University Press.

Dancy, J. (2004). *Ethics without principles*. New York: Oxford University Press.

Darley, J. M., & Schultz, T. R. (1990). Moral rules: Their content and acquisition. *Annual Review of Psychology, 41*, 525–556. doi:10.1146/annurev.ps.41.020190.002521.

D'Arms, J. (2005). Two arguments for sentimentalism. *Philosophical Issues, 15*(1), 1–21. doi:10.1111/j.1533-6077.2005.00050.x.

D'Arms, J., & Jacobson, D. (2000a). Sentiment and value. *Ethics, 110*(4), 722–748. doi:10.1086/233371.

D'Arms, J., & Jacobson, D. (2000b). The moralistic fallacy: On the "appropriateness" of emotions. *Philosophy and Phenomenological Research, 61*(1), 65–90.

Darwall, S. (2006). *The second-person standpoint: Morality, respect, and accountability*. Cambridge, MA: Harvard University Press.

Darwall, S., Gibbard, A., & Railton, P. (1992). Toward fin de siècle ethics: Some trends. *Philosophical Review, 101*(1), 115–189.

Davidson, D. (1989). A coherence theory of truth and knowledge. *Truth and interpretation: Perspectives on the philosophy of Donald Davidson* (pp. 307–319). E. LePore, ed. New York: Blackwell.

Dean, R. (2010). Does neuroscience undermine deontological theory? *Neuroethics, 3(1)*, 43–60. doi:10.1007/s12152-009-9052-x.

Decety, J., & Cowell, J. M. (2014). The complex relation between morality and empathy. *Trends in Cognitive Sciences, 18*(7), 337–339. doi:10.1016/j.tics.2014.04.008.

Deigh, J. (1994). Cognitivism in the theory of emotions. *Ethics, 104*(4), 824–854. doi:10.1086/293657.

de Lazari-Radek, K., & Singer, P. (2012). The objectivity of ethics and the unity of practical reason. *Ethics, 123(1)*, 9–31. doi:10.1086/667837.

Deonna, J. A. (2006). Emotion, perception and perspective. *Dialectica, 60*(1), 29–46. doi:10.1111/j.1746-8361.2005.01031.x.

de Rose, K. (1995). Solving the skeptical problem. *Philosophical Review, 104(1)*, 1–52. doi:10.2307/2186011.

de Sousa, R. (1987). *The rationality of emotion*. Cambridge, MA: MIT Press.

de Sousa, R. (2001). Moral emotions. *Ethical Theory and Moral Practice, 4*(2), 109–126. doi:10.1023/A:1011434921610.

de Waal, F. (2006). *Primates and philosophers: How morality evolved*. Princeton, NJ: Princeton University Press.

Dijksterhuis, A. (2004). Think different: The merits of unconscious thought in preference development and decision making. *Journal of Personality and Social Psychology, 87*(5), 586–598. doi:10.1037/0022-3514.87.5.586.

Dijksterhuis, A. (2006). On the benefits of thinking unconsciously: Unconscious thought can increase post-choice satisfaction. *Journal of Experimental Social Psychology, 42(5)*, 627–631. doi:10.1016/j.jesp.2005.10.008.

Ditto, P. H., Pizarro, D. A., & Tannenbaum, D. (2009). Motivated moral reasoning. In D. M. Bartels, C. W. Bauman, L. J. Skitka, & D. L. Medin (Eds.), *The psychology of learning and motivation*, vol. 50 (pp. 307–338). Burlington: Academic Press.

Döring, S. (2007). Seeing what to do: Affective perception and rational motivation. *Dialectica, 61*(3), 363–394. doi:10.1111/j.1746-8361.2007.01105.x.

Doris, J. M. (2015). *Talking to our selves: Reflection, ignorance, and agency*. Oxford: Oxford University Press.

Doris, J. M., & Stich, S. (2005). As a matter of fact: Empirical perspectives on ethics. In F. Jackson and M. Smith (Eds.), *The Oxford handbook of contemporary philosophy* (pp. 114–152). New York: Oxford University Press.

Dreyfus, H. L., & Dreyfus, S. E. (1986). *Mind over machine: The power of human intuitive expertise in the era of the computer*. New York: Free Press.

Dreyfus, H. L., & Dreyfus, S. E. (1991). Towards a phenomenology of ethical expertise. *Human Studies, 14(4)*, 229–250. doi:10.1007/BF02205607.

Duhigg, C. (2012). *The power of habit: Why we do what we do and how to change*. London: William Heinemann.

Duke, A. A., & Bègue, L. (2015). The drunk utilitarian: Blood alcohol concentration predicts utilitarian responses in moral dilemmas. *Cognition, 134*, 121–127. doi:10.1016/j.cognition.2014.09.006.

Dunning, D. (1999). A newer look: Motivated social cognition and the schematic representation of social concepts. *Psychological Inquiry, 10*(1), 1–11. doi:10.1207/s15327965pli1001_1.

Earp, B. D., Everett, J. A. C., Madva, E. N., & Hamlin, J. K. (2014). Out, damned spot: Can the "Macbeth Effect" be replicated? *Basic and Applied Social Psychology, 36(1)*, 91–98. doi:10.1080/01973533.2013.856792.

Egan, A. (2007). Quasi-realism and fundamental moral error. *Australasian Journal of Philosophy, 85*(2), 205–219. doi:10.1080/00048400701342988.

Elias, N. (1994). *The civilizing process: Sociogenetic and psychogenetic investigations*. Malden, MA: Blackwell.

Enoch, D. (2009). How is moral disagreement a problem for moral realism? *Journal of Ethics, 13(1)*, 15–50. doi:10.1007/s10892-008-9041-z.

Eskine, K. J., Kacinik, N. A., & Prinz, J. J. (2011). A bad taste in the mouth: Gustatory disgust influences moral judgment. *Psychological Science, 22*(3), 295–299. doi:10.1177/0956797611398497.

Evans, J. St. B. T. (2003). In two minds: Dual-process accounts of reasoning. *Trends in Cognitive Sciences, 7*(10), 454–459. doi:10.1016/j.tics.2003.08.012.

Evans, J. St. B. T. (2008). Dual-processing accounts of reasoning, judgment, and social cognition. *Annual Review of Psychology, 59*, 255–278. doi:10.1146/annurev.psych.59.103006.093629.

Evans, J. St. B. T., & Stanovich, K. E. (2013). Dual-process theories of higher cognition: Advancing the debate. *Perspectives on Psychological Science, 8*(3), 223–241. doi:10.1177/1745691612460685.

Feinberg, M., Willer, R., Antonenko, O., & John, O. P. (2012). Liberating reason from the passions: Overriding intuitionist moral judgments through emotion reappraisal. *Psychological Science, 23*(7), 788–795. doi:10.1177/0956797611434747.

Fine, C. (2006). Is the emotional dog wagging its rational tail, or chasing it? Reason in moral judgment. *Philosophical Explorations, 9*(1), 83–98. doi:10.1080/13869790500492680.

Fitzpatrick, S. (2014). Moral realism, moral disagreement, and moral psychology. *Philosophical Papers, 43*(2), 161–190. doi:10.1080/05568641.2014.932953.

Flanagan, O. (1991). *Varieties of moral personality: Ethics and psychological realism.* Cambridge, MA: Harvard University Press.

Fodor, J. A. (1983). *The modularity of mind.* Cambridge, MA: MIT Press.

Foot, P. (1967). The problem of abortion and the doctrine of double effect. *Oxford Review, 5,* 5–15.

Foucault, M. (1977). *Discipline and punish: The birth of the prison.* New York: Pantheon.

Freeman, W. (1950). Psychosurgery. *Journal of the National Medical Association, 42,* 206–209.

Frith, U. (1989). *Autism: Explaining the enigma.* Oxford: Blackwell.

Fumagalli, M., & Priori, A. (2012). Functional and clinical neuroanatomy of morality. *Brain, 135*(7), 2006–2021. doi:10.1093/brain/awr334.

Gallo, I. S., Keil, A., McCulloch, K. C., Rockstroh, B., & Gollwitzer, P. M. (2009). Strategic automation of emotion regulation. *Journal of Personality and Social Psychology, 96*(1), 11–31. doi:10.1037/a0013460.

Garfinkel, H. (1964). Studies of the routine grounds of everyday activities. *Social Problems, 11*(3), 225–250.

Gazzaniga, M. S. (1983). Right hemisphere language following brain bisection: A twenty year perspective. *American Psychologist, 38*(5), 525–537. doi:10.1037/0003-066X.38.5.525.

Geach, P. T. (1965). Assertion. *Philosophical Review, 74*(4), 449–465. doi:10.2307/2183123.

Gehlen, A. (1940). *Der Mensch: Seine Natur und seine Stellung in der Welt.* Berlin: Junker and Dünnhaupt.

Gerrans, P., & Kennett, J. (2006). Introduction: Is cognitive penetrability the mark of the moral? *Philosophical Explorations, 9*(1), 3–12. doi:10.1080/13869790500492284.

Gerrans, P., & Kennett, J. (2010). Neurosentimentalism and moral agency. *Mind, 119*(475), 585–614. doi:10.1093/mind/fzq037.

Gert, J. (2002). Avoiding the conditional fallacy. *Philosophical Quarterly, 52*(206), 88–95. doi:10.1111/1467-9213.00255.

Gert, J. (2003). Brute rationality. *Noûs, 37*(3), 417–446.

Gert, J. (2015). Disgust, moral disgust, and morality. *Journal of Moral Philosophy, 12*(1), 33–54. doi:10.1163/17455243-4681048.

Gettier, E. L. (1963). Is justified true belief knowledge? *Analysis, 23(6)*, 121–123.

Gibbard, A. (1986). An expressivistic theory of normative discourse. *Ethics, 96*(3), 472–485. doi:10.1086/292770.

Gibbard, A. (1990). *Wise choices, apt feelings: A theory of normative judgment.* New York: Oxford University Press.

Gibbard, A. (1992). Moral concepts: Substance and sentiment. *Philosophical Perspectives, 6*, 199–221. doi:10.2307/2214245.

Giddens, A. (1984). *The constitution of society: Outline of the theory of structuration.* Cambridge: Polity Press.

Gigerenzer, G. (2007). *Gut feelings: Short cuts to better decision making.* London: Penguin Books.

Gigerenzer, G. (2008). Moral intuition = fast and frugal heuristics? In W. Sinnott-Armstrong (ed.), *Moral psychology*, Vol. 2: *The cognitive science of morality: Intuition and diversity* (pp. 1–26). Cambridge, MA: MIT Press.

Gilligan, C. (1982). *In a different voice: Psychological theory and women's development.* Cambridge, MA: Harvard University Press.

Gilovich, T., Griffin, D., & Kahneman, D. (Eds.). (2002). *Heuristics and biases: The psychology of intuitive judgment.* New York: Cambridge University Press.

Goldie, P. (2000). *The emotions: A philosophical exploration.* New York: Oxford University Press.

Goldie, P. (2004a). *On personality*. London: Routledge.

Goldie, P. (2004b). Emotion, reason, and virtue. In D. Evans & P. Cruse (Eds.), *Emotion, evolution, and rationality* (pp. 249–267). Oxford: Oxford University Press.

Goldie, P. (2004c). Emotion, feeling, and knowledge of the world. In R. Solomon (Ed.), *Thinking about feeling: Contemporary philosophers on emotion.* Oxford: Oxford University Press.

Goldie, P. (2007). Seeing what is the kind of thing to do: Perception and emotion in morality. *Dialectica, 61*(3), 347–361. doi:10.1111/j.1746-8361.2007.01107.x.

Goldman, A. (1967). A causal theory of knowing. *Journal of Philosophy, 64*(12), 357–372. doi:10.2307/2024268.

Goldman, A. (1976). Discrimination and perceptual knowledge. *Journal of Philosophy, 73(20)*, 771–791. doi:10.2307/2025679.

Gollwitzer, P. (1999). Implementation intentions: Strong effects of simple plans. *American Psychologist, 54*(7), 493–503. doi:10.1037//0003-066X.54.7.493.

Graham, J., Haidt, J., & Nosek, B. A. (2009). Liberals and conservatives rely on different sets of moral foundations. *Journal of Personality and Social Psychology, 96*(5), 1029–1046. doi:10.1037/a0015141.

Gray, K., Schein, C., & Ward, A. F. (2014). The myth of harmless wrongs in moral cognition: Automatic dyadic completion from sin to suffering. *Journal of Experimental Psychology, 143*(4), 1600–1615. doi:10.1037/a0036149.

Gray, K., & Wegner, D. (2009). Moral typecasting: Divergent perceptions of moral agents and moral patients. *Journal of Personality and Social Psychology, 96*(3), 505–520. doi:10.1037/a0013748.

Graybiel, A. M. (2008). Habits, rituals, and the evaluative brain. *Annual Review of Neuroscience, 31*, 359–387. doi:10.1146/annurev.neuro.29.051605.112851.

Greene, J. D. (2003). From neural "is" to moral "ought": What are the moral implications of neuroscientific moral psychology? *Nature Reviews Neuroscience, 4*, 847–850. doi:10.1038/nrn1224.

Greene, J. D. (2008). The secret joke of Kant's soul. In W. Sinnott-Armstrong (Ed.), *Moral psychology*, Vol. 3: *The neuroscience of morality: Emotion, brain disorders, and development*. Cambridge, MA: MIT Press.

Greene, J. D. (2009). Dual-process morality and the personal/impersonal distinction: A reply to McGuire, Langdon, Coltheart, and Mackenzie. *Journal of Experimental Social Psychology, 45*(3), 581–584. doi:10.1016/j.jesp.2009.01.003.

Greene, J. D., Cushman, F., Stewart, L. E., Lowenberg, K., Nystrom, L. E., & Cohen, J. D. (2009). Pushing moral buttons: The interaction between personal force and intention in moral judgment. *Cognition, 111*(3), 364–371. doi:10.1016/j.cognition.2009.02.001.

Greene, J. D., Morelli, S. A., Lowenberg, K., Nystrom, L. E., & Cohen, J. D. (2008). Cognitive load selectively interferes with utilitarian moral judgment. *Cognition, 107(3)*, 1144–1154. doi:10.1016/j.cognition.2007.11.004.

Greene, J. D., Nystrom, L. E., Engell, A. D., Darley, J. M., & Cohen, J. D. (2004). The neural bases of cognitive conflict and control in moral judgment. *Neuron, 44(2)*, 389–400. doi:10.1016/j.neuron.2004.09.027.

Greene, J. D., Sommerville, B. D., Nystrom, L. E., Darley, J. M., & Cohen, J. D. (2001). An fMRI investigation of emotional engagement in moral judgment. *Science, 293*(5537), 2105–2108. doi:10.1126/science.1062872.

Greenspan, P. (1988). *Emotions and reasons: An inquiry into emotional justification.* New York: Routledge, Chapman & Hall.

Greenspan, P. (2015). Confabulating the truth: In defense of "defensive" moral reasoning. *Journal of Ethics, 19*(2), 105–123. doi:10.1007/s10892-015-9193-6.

Gross, J. J. (2002). Emotion regulation: Affective, cognitive, and social consequences. *Psychophysiology, 39*(*3*), 281–291. doi:10.1017/S0048577201393198.

Gunther, Y. (2003). Emotions and force. In Y. Gunther (Ed.), *Essays on nonconceptual content* (pp. 279–288). Cambridge, MA: MIT Press.

Haack, S. (1993). Double-aspect foundherentism: A new theory of empirical justification. *Philosophy and Phenomenological Research, 53*(1), 113–128. doi:10.2307/2108056.

Haidt, J. (2001). The emotional dog and its rational tail: A social intuitionist approach to moral judgement. *Psychological Review, 108*(*4*), 814–834. doi:10.1037/0033-295X.108.4.814.

Haidt, J. (2003). The moral emotions. In R. J. Davidson, K. R. Scherer, & H. Hill Goldsmith (Eds.), *Handbook of the affective sciences* (pp. 852–870). Oxford: Oxford University Press.

Haidt, J. (2004). The emotional dog gets mistaken for a possum. *Review of General Psychology, 8*(4), 283–290. doi:10.1037/1089-2680.8.4.283.

Haidt, J. (2007). The new synthesis in moral psychology. *Science, 316*(*5827*), 998–1001. doi:10.1126/science.1137651.

Haidt, J. (2012). *The righteous mind: Why good people are divided by religion and politics.* London: Penguin.

Haidt, J., & Hersh, M. (2001). Sexual morality: The cultures and emotions of conservatives and liberals. *Journal of Applied Social Psychology, 31*(*1*), 191–221. doi:10.1111/j.1559-1816.2001.tb02489.x.

Haidt, J., & Bjorklund, F. (2008). Social intuitionists answer six questions about moral psychology. In W. Sinnott-Armstrong (Ed.), *Moral psychology*, Vol. 2: *The cognitive science of morality: intuition and diversity* (pp. 181–217). Cambridge, MA: MIT Press.

Haidt, J., Björklund, F., & Murphy, S. (2000). Moral dumbfounding: When intuition finds no reason. Unpublished manuscript, University of Virginia.

Haidt, J., & Kesebir, S. (2010). Morality. In S. Fiske, D. Gilbert, & G. Lindzey (Eds.), *Handbook of social psychology* (pp. 797–832). Hoboken, NJ: Wiley.

Haidt, J., Koller, S., & Dias, M. G. (1993). Affect, culture, and morality, or is it wrong to eat your dog? *Journal of Personality and Social Psychology, 65*(4), 613–628. doi:10.1037/0022-3514.65.4.613.

Haidt, J., Rozin, P., McCauley, C., & Imada, S. (1997). Body, psyche, and culture: The relationship between disgust and morality. *Psychology and Developing Societies, 9*(1), 107–131. doi:10.1177/097133369700900105.

Hall, L., Johansson, P., & Strandberg, T. (2012). Lifting the veil of morality: Choice blindness and attitude reversals on a self-transforming survey. *PLoS One, 7*(9), e45457. doi:10.1371/journal.pone.0045457.

Hare, R. D. (1993). *Without conscience: The disturbing world of the psychopaths among us*. New York: Pocket Books.

Hare, R. D. (1999). *Without conscience: The disturbing world of the psychopaths among us*. New York: Guilford Press.

Harman, G. (1999). Moral philosophy meets social psychology: Virtue ethics and the fundamental attribution error. *Proceedings of the Aristotelian Society, 99*, 315–331.

Harman, G. (2007). Ethics and observation. In R. Shafer-Landau (Ed.), *Ethical theory: An anthology* (pp. 36–41). Malden, MA: Blackwell.

Hauser, M. D. (2006). *Moral minds: How nature designed our universal sense of right and wrong*. New York: Ecco.

Hauser, M., Cushman, F., Young, L., Jin, R. K.-X., & Mikhail, J. (2007). A dissociation between moral judgments and justifications. *Mind & Language, 22*(1), 1–21. doi:10.1111/j.1468-0017.2006.00297.x.

Heath, J., & Hardy-Vallée, B. (2015). Why do people behave immorally when drunk? *Philosophical Explorations, 18*(3), 310–329. doi:10.1080/13869795.2015.996585.

Hegel, G. W. F. (1986). *Phänomenologie des Geistes*. Frankfurt am Main: Suhrkamp.

Hegel, G. W. F. (1988). *Introduction to the philosophy of history*. Indianapolis, IN: Hackett.

Heidegger, M. (1927). *Sein und Zeit*. Tübingen: Mohr.

Helm, B. W. (2001). *Emotional reason: Deliberation, motivation, and the nature of value*. Cambridge, MA: Cambridge University Press.

Henson, R. (2005). What can functional neuroimaging tell the experimental psychologist? *Quarterly Journal of Experimental Psychology, 58*(2), 193–233. doi:10.1080/02724980443000502.

Henrich, J., Heine, S. J., & Norenzayan, A. (2010). The weirdest people in the world? *Behavioral and Brain Sciences, 33*(2–3), 61–83. doi:10.1017/S0140525X0999152X.

Hindriks, F. (2015). How does reasoning (fail to) contribute to moral judgment? Dumbfounding and disengagement. *Ethical Theory and Moral Practice, 18*(2), 237–250. doi:10.1007/s10677-015-9575-7.

Hirstein, W. (2005). *Brain fiction: Self-deception and the riddle of confabulation.* Cambridge, MA: MIT Press.

Hogarth, R. M. (2001). *Educating intuition.* Chicago: University of Chicago Press.

Horgan, T., & Timmons, M. (2007). Morphological rationalism and the psychology of moral judgment. *Ethical Theory and Moral Practice, 10(3)*, 279–295. doi:10.1007/s10677-007-9068-4.

Huemer, M. (2005). *Ethical intuitionism.* New York: Palgrave Macmillan.

Hume, D. (1975). *A treatise of human nature.* Oxford: Clarendon.

Inbar, Y., Pizarro, D. A., & Bloom, P. (2009). Conservatives are more easily disgusted than liberals. *Cognition and Emotion, 23*(4), 714–725. doi:10.1080/02699930802110007.

Jacobson, D. (2005). Seeing by feeling: virtues, skills, and moral perception. *Ethical Theory and Moral Practice, 8(4)*, 387–409. doi:10.1007/s10677-005-8837-1.

Jacobson, D. (2008). Does social intuitionism flatter morality or challenge it? In W. Sinnott-Armstrong (Ed.), *Moral psychology*, Vol. 2: *The cognitive science of morality: intuition and diversity* (pp. 219–233). Cambridge, MA: MIT Press.

Jacobson, D. (2012). Moral dumbfounding and moral stupefaction. *Oxford Studies in Normative Ethics*, Vol. 2. New York: Oxford University Press.

James, W. (1884). What is an emotion? *Mind, 9*, 188–205.

James, W. ([1890] 1950). *The principles of psychology.* New York: Dover.

Johansson, P., Hall, L., Sikström, S., & Olsson, A. (2005). Failure to detect mismatches between intention and outcome in a simple decision task. *Science, 310(5745)*, 116–119. doi:10.1126/science.1111709.

Johnson, D. J., Cheung, F., & Donnellan, M. B. (2014). Does cleanliness influence moral judgments? *Social Psychology, 45(3)*, 209–215. doi:10.1027/1864-9335/a000186.

Johnson, R. (1999). Internal reasons and the conditional fallacy. *Philosophical Quarterly, 49*(149), 53–71. doi:10.1111/1467-9213.00128.

Jones, A. P., Happé, F. G., Gilbert, F., Burnett, S., & Viding, E. (2010). Feeling, caring, knowing: Different types of empathy deficit in boys with psychopathic tendencies and autism spectrum disorder. *Journal of Child Psychology and Psychiatry, and Allied Disciplines, 51*(11), 1188–1197. doi:10.1111/j.1469-7610.2010.02280.x.

Jones, K. (2006a). Metaethics and emotions research: A response to Prinz. *Philosophical Explorations, 9(1)*, 45–53. doi:10.1080/13869790500492508.

Jones, K. (2006b). Quick and smart? Modularity and the pro-emotion consensus. *Canadian Journal of Philosophy* 36(Suppl. 32): 3–27.

Joyce, R. (2007). *The evolution of morality*. Cambridge, MA: MIT Press.

Kahan, D. M. (1999). The progressive appropriation of disgust. In S. A. Bandes (Ed.), *The passions of law* (pp. 63–79). New York: New York University Press.

Kahane, G. (2012). On the wrong track: Process and content in moral psychology. *Mind & Language, 27*(5), 519–545. doi:10.1111/mila.12001.

Kahane, G. (2013). Must metaethical realism make a semantic claim? *Journal of Moral Philosophy, 10*(2), 148–178. doi:10.1163/174552412X628869.

Kahane, G., Everett, J. A. C., Earp, B. D., Farias, M., & Savulescu, J. (2015). "Utilitarian" judgments in sacrificial moral dilemmas do not reflect impartial concern for the greater good. *Cognition, 134*, 193–209. doi:10.1016/j.cognition.2014.10.005.

Kahane, G., & Shackel, N. (2010). Methodological issues in the neuroscience of moral judgment. *Mind & Language, 25*(5), 561–582.

Kahane, G., Wiech, K., Shackel, N., Farias, M., Savulescu, J., & Tracey, I. (2012). The neural basis of intuitive and counterintuitive moral judgment. *Social Cognitive and Affective Neuroscience, 7*(4), 393–402. doi:10.1093/scan/nsr005.

Kahneman, D. (2003). A perspective on judgment and choice: Mapping bounded rationality. *American Psychologist, 58*(9), 697–720. doi:10.1037/0003-066X.58.9.697.

Kahneman, D., & Frederick, S. (2002). Representativeness revisited: Attribute substitution in intuitive judgment. In T. Gilovich, D. Griffin and D. Kahneman (eds.), *Heuristics and biases: The psychology of intuitive judgment* (pp. 49–81). New York: Cambridge University Press.

Kahneman, D., Slovic, P., & Tversky, A. (Eds.). (1982). *Judgment under uncertainty: Heuristics and biases*. Cambridge: Cambridge University Press.

Kamm, F. M. (2007). *Intricate ethics: Rights, responsibilities, and permissible harm*. New York: Oxford University Press.

Kamm, F. M. (2009). Neurocience and moral reasoning: A note on recent research. *Philosophy & Public Affairs, 37*(4), 330–345. doi:10.1111/j.1088-4963.2009.01165.x.

Kass, L. R. (1997). The wisdom of repugnance. *New Republic, 216*(22), 17–26.

Kelly, D. (2011). *Yuck! The nature and moral significance of disgust*. Cambridge, MA: MIT Press.

Kennett, J. (2006). Do psychopaths really threaten moral rationalism? *Philosophical Explorations, 9*(1), 69–82. doi:10.1080/13869790500492524.

Kennett, J. (2011). Imagining reasons. *Southern Journal of Philosophy, 49*(Suppl. 1), 181–192. doi:10.1111/j.2041-6962.2011.00066.x.

Kennett, J., & Fine, C. (2008). Internalism and the evidence from psychopaths and "acquired sociopaths." In W. Sinnott-Armstrong (ed.), *Moral psychology*, Vol. 3: *The neuroscience of morality: Emotion, brain disorders, and development* (pp. 173–191). Cambridge, MA: MIT Press.

Kennett, J., & Fine, C. (2009). Will the real moral judgment please stand up? The implications of social intuitionist models of cognition for meta-ethics and moral psychology. *Ethical Theory and Moral Practice, 12(1)*, 77–96. doi:10.1007/s10677-008-9136-4.

Kenny, A. (1963). *Action, emotion, and will*. London: Routledge & Kegan Paul.

Keren, G., & Schul, Y. (2009). Two is not always better than one: A critical evaluation of two-system theories. *Perspectives on Psychological Science, 4*(6), 533–550. doi:10.1111/j.1745-6924.2009.01164.x.

Klein, C. (2010). The dual track theory of moral decision-making: A critique of the neuroimaging evidence. *Neuroethics, 4*(2), 143–162. doi:10.1007/s12152-010-9077-1.

Klingberg, T. (2009). *The overflowing brain: Information overload and the limits of working memory*. New York: Oxford University Press.

Knapp, C. (2003). De-moralizing disgustingness. *Philosophy and Phenomenological Research, 66*(2), 253–278. doi:10.1111/j.1933-1592.2003.tb00261.x.

Knobe, J. (2003). Intentional action and side effects in ordinary language. *Analysis, 63(3)*, 190–193. doi:10.1111/1467-8284.00419.

Knobe, J., & Nichols, S. (Eds.). (2008). *Experimental philosophy*. New York: Oxford University Press.

Kober, H., Barrett, L. F., Joseph, J., Bliss-Moreau, E., Lindquist, K., & Wager, T. D. (2008). Functional grouping and cortical–subcortical interactions in emotion: A meta-analysis of neuroimaging studies. *NeuroImage, 42(2)*, 998–1031. doi:10.1016/j.neuroimage.2008.03.059.

Koenigs, M., & Tranel, D. (2007). Irrational economic decision-making after ventromedial prefrontal damage: Evidence from the ultimatum game. *Journal of Neuroscience, 27*(4), 951–956. doi:10.1523/JNEUROSCI.4606-06.2007.

Koenigs, M., Young, L., Adolphs, R., Tranel, D., Cushman, F., Hauser, M., & Damasio, A. (2007). Damages to the prefrontal cortex increases utilitarian moral judgments. *Nature, 446*(7138), 908–911. doi:10.1038/nature05631.

Kohlberg, L. (1969). Stage and sequence: The cognitive-developmental approach to socialization. In D. A. Goslin (Ed.), *Handbook of socialization theory and research* (pp. 347–480). Chicago: Rand McNally.

Kornblith, H. (2010). What reflective endorsement cannot do. *Philosophy and Phenomenological Research, 80*(1), 1–19. doi:10.1111/j.1933-1592.2009.00307.x.

Korsgaard, C. (1996). *The sources of normativity*. New York: Cambridge University Press.

Korsgaard, C. (2009). *Self-constitution: Agency, identity, and integrity*. Oxford: Oxford University Press.

Krause, S. (2008). *Civil passions: Moral sentiment and democratic deliberation*. Princeton, NJ: Princeton University Press.

Kundu, P., & Cummins, D. D. (2013). Morality and conformity: The Asch paradigm applied to moral decisions. *Social Influence, 8*(4), 268–279. doi:10.1080/15534510.2012.727767.

La Caze, M. (2001). Envy and resentment. *Philosophical Explorations, 4*(1), 31–45. doi:10.1080/13869790108523341.

Lally, P., van Jaarsveld, C. H. M., Potts, H. W. W., & Wardle, J. (2010). How are habits formed: Modelling habit formation in the real world. *European Journal of Social Psychology, 40(6)*, 998–1009. doi:10.1002/ejsp.674.

Landy, J. F., & Goodwin, G. P. (2015). Does incidental disgust amplify moral judgment? A meta-analytic review of experimental evidence. *Perspectives on Psychological Science, 10*(4), 518–536. doi:10.1177/1745691615583128.

Langer, E., Blank, A., & Chanowitz, B. (1978). The mindlessness of ostensibly thoughtful action: The role of "placebic" information in interpersonal interaction. *Journal of Personality and Social Psychology, 36*(6), 635–642. doi:10.1037/0022-3514.36.6.635.

Lapsley, D. K. and Hill, P. L. (2008). On dual processing and heuristic approaches to moral cognition. *Journal of Moral Education, 37*(3), 313–332.

Lapsley, D. K., & Narvaez, D. (2004). A social-cognitive approach to the moral personality. In D. K. Lapsley & D. Narvaez (Eds.), *Moral development, self, and identity* (pp. 189–212). Mahwah, NJ: Erlbaum.

Leff, L. J. and J. L. Simmons (2001). *The dame in the kimono: Hollywood, censorship, and the production code* (2nd ed.). Lexington: University Press of Kentucky.

Lehrer, K. (2000). *Theory of knowledge*. Boulder, CO: Westview.

Leplin, J. (2009). *A theory of epistemic justification*. New York: Springer.

Leslie, A. M., Mallon, R., & DiCorcia, J. A. (2006). Transgressors, victims, and cry babies: Is basic moral judgment spared in autism? *Social Neuroscience, 1*(3–4), 270–283. doi:10.1080/17470910600992197.

Levy, N. (2007). *Neuroethics: Challenges for the 21st century*. Cambridge: Cambridge University Press.

Levy, N. (2014). Psychopaths and blame: The argument from content. *Philosophical Psychology, 27*(3), 351–367. doi:10.1080/09515089.2012.729485.

Lewis, D. (1996). Elusive knowledge. *Australasian Journal of Philosophy, 74(4)*, 549–567. doi:10.1080/00048409612347521.

Liao, M. (2011). Bias and reasoning: Haidt's theory of moral judgment. In T. Brooks (Ed.), *New waves in ethics* (pp. 108–128). London: Palgrave MacMillan.

Lillehammer, H. (2007). *Companions in guilt: Arguments for ethical objectivity*. New York: Palgrave Macmillan.

Logothetis, N. K. (2008). What we can do and what we cannot do with fMRI. *Nature, 453(7197)*, 869–878. doi:10.1038/nature06976.

Louise, J. (2009). Correct responses and the priority of the normative. *Ethical Theory and Moral Practice, 12(4)*, 345–364. doi:10.1007/s10677-009-9177-3.

Lovibond, S. (2002). *Ethical formation*. Cambridge, MA: Harvard University Press.

Machery, E. (2014). In defense of reverse inference. *British Journal for the Philosophy of Science, 65*(2), 251–267. doi:10.1093/bjps/axs044.

Mackie, J. L. (1977). *Ethics: Inventing right and wrong*. London: Penguin.

Maddock, R. J. (1999). The retrosplenial cortex and emotion: New insights from functional neuroimaging of the human brain. *Trends in Neurosciences, 22*(7), 310–316. doi:10.1016/S0166-2236(98)01374-5.

Maibom, H. (2005). Moral unreason: The case of psychopathy. *Mind & Language, 20*(2), 237–257. doi:10.1111/j.0268-1064.2005.00284.x.

Maibom, H. (2008). The mad, the bad, and the psychopath. *Neuroethics, 1(3)*, 167–184. doi:10.1007/s12152-008-9013-9.

Maibom, H. (2010). What experimental evidence shows us about the role of emotions in moral judgement. *Philosophy Compass, 5(11)*, 999–1012. doi:10.1111/j.1747-9991.2010.00341.x.

Mason, M. (2003). Contempt as a moral attitude. *Ethics, 113(2)*, 234–272. doi:10.1086/342860.

May, J. (2014). Does disgust influence moral judgment? *Australasian Journal of Philosophy, 92*(1), 125–141. doi:10.1080/00048402.2013.797476.

McDowell, J. ([1978] 1998). Are moral requirements hypothetical imperatives? In *Mind, value, and reality* (pp. 77–94). Cambridge, MA: Harvard University Press.

McDowell, J. (1985). Values and secondary qualities. In T. Honderich (Ed.), *Morality and objectivity* (pp. 110–129). Boston: Routledge & Kegan Paul.

McDowell, J. (1994). *Mind and world*. Cambridge, MA: Harvard University Press.

McDowell, J. (1998). *Projection and truth in ethics: Mind, value, and reality* (pp. 151–167). Cambridge, MA: Harvard University Press.

McGeer, V. (2008). Varieties of moral agency: Lessons from autism (and psychopathy). In W. Sinnott-Armstrong (ed.), *Moral psychology*, Vol. 3: *The neuroscience of morality: Emotion, brain disorders, and development* (pp. 227–259). Cambridge, MA: MIT Press.

McGuire, J., Langdon, R., Coltheart, M., & Mackenzie, C. (2009). A reanalysis of the personal/impersonal distinction in moral psychology research. *Journal of Experimental Social Psychology, 45*(3), 577–580. doi:10.1016/j.jesp.2009.01.002.

McKeever, S., & Ridge, M. (2006). *Principled ethics: Generalism as a regulative ideal*. Oxford: Oxford University Press.

Mendez, M., Anderson, E., & Shapira, J. S. (2005). An investigation of moral judgement in frontotemporal dementia. *Cognitive and Behavioral Neurology, 18*(4), 193–197. doi:10.1097/01.wnn.0000191292.17964.bb.

Mercier, H. (2011). What is good moral reasoning? *Mind & Society, 10*(2), 131–148. doi:10.1007/s11299-011-0085-6.

Mercier, H., & Sperber, D. (2011). Why do humans reason? Arguments for an argumentative theory. *Behavioral and Brain Sciences, 34*(2), 57–111. doi:10.1017/S0140525X10000968.

Mikhail, J. (2007). Universal moral grammar: Theory, evidence, and the future. *Trends in Cognitive Sciences, 11*(4), 143–152. doi:10.1016/j.tics.2006.12.007.

Miller, A. (2003). *An introduction to contemporary metaethics*. Cambridge: Polity Press.

Miller, G. (2008). Growing pains for fMRI. *Science, 320*, 1412–1414.

Millgram, E. (1999). Moral values and secondary qualities. *American Philosophical Quarterly, 36*(3), 253–255.

Moll, A. (1889). *Hypnotism*. London: Walter Scott.

Moll, J., & de Oliveira-Souza, R. (2007). Moral judgments, emotions, and the utilitarian brain. *Trends in Cognitive Sciences, 11*(8), 319–321. doi:10.1016/j.tics.2007.06.001.

Moll, J., de Oliveira-Souza, R., & Eslinger, P. J. (2003). Morals and the human brain: A working model. *Neuroreport, 14*(3), 299–305. doi:10.1097/00001756-200303030-00001.

Moll, J., de Oliveira-Souza, R., & Zahn, R. (2008). The neural basis of moral cognition: Sentiments, concepts, and values. *Annals of the New York Academy of Sciences, 1124*, 161–180. doi:10.1196/annals.1440.005.

Monteith, M., & Mark, A. (2005). Changing one's prejudiced ways: Awareness, affect, and self-regulation. *European Review of Social Psychology, 16*(1), 113–154. doi:10.1080/10463280500229882.

Moody-Adams, M. (1997). *Fieldwork in familiar places: Morality, culture, and philosophy*. Cambridge, MA: Harvard University Press.

Moore, A. B., Clark, B. A., & Kane, M. J. (2008). Who shalt not kill? Individual differences in working memory capacity, executive control, and moral judgment. *Psychological Science, 19*(6), 549–557. doi:10.1111/j.1467-9280.2008.02122.x.

Morton, A. (2004). Epistemic virtues, metavirtues, and computational complexity. *Noûs, 38*(3), 481–502. doi:10.1111/j.0029-4624.2004.00479.x.

Murphy, J. (1972). Moral death: A Kantian essay on psychopathy. *Ethics, 82(4)*, 284–298. doi:10.1086/291852.

Musschenga, A. (2008). Moral judgement and moral reasoning: A critique of Jonathan Haidt. In M. Düwell, C. Rehmann-Sutter, & D. Mieth (Eds.), *The contingent nature of human life: Bioethics and the limits of human existence* (pp. 131–141). Dordrecht: Springer. doi:10.1007/978-1-4020-6764-8_12.

Musschenga, A. (2009). Moral intuitions, moral expertise and moral reasoning. *Journal of Philosophy of Education, 43*(4), 597–613. doi:10.1111/j.1467-9752.2009.00707.x.

Nichols, S. (2004). *Sentimental rules: On the natural foundations of moral judgment*. New York: Oxford University Press.

Nichols, S. (2008). Sentimentalism naturalized. In W. Sinnott-Armstrong (Ed.), *Moral psychology*, Vol. 2: *The cognitive science of morality: Intuition and diversity* (pp. 255–275). Cambridge, MA: MIT Press.

Nichols, S., Kumar, S., & Lopez, T. (forthcoming). Rational learners and moral rules. *Mind & Language*.

Nichols, S., & Mallon, R. (2006). Moral dilemmas and moral rules. *Cognition, 100(3)*, 530–542. doi:10.1016/j.cognition.2005.07.005.

Nietzsche, F. ([1886] 1966). *Beyond good and evil: Prelude to a philosophy of the future*. New York: Random House.

Nisbett, R. E., & Wilson, T. D. (1977). Telling more than we can know: Verbal reports on mental processes. *Psychological Review, 84*(3), 231–259. doi:10.1037/0033-295X.84.3.231.

Nisbett, R. E., & Wilson, T. D. (1978). The accuracy of verbal reports about the effects of stimuli and behavior. *Social Psychology, 41*(2), 118–131. doi:10.2307/3033572.

Nozick, R. (1981). *Philosophical explanations*. Cambridge, MA: Belknap Press of Harvard University Press.

Nozick, R. (1993). *The nature of rationality*. Princeton, NJ: Princeton University Press.Nucci, L. (1985). Children's conceptions of morality, social conventions and religious prescription. In C. Harding (Ed.), *Moral dilemmas: Philosophical and psychological reconsiderations of moral reasoning* (pp. 137–174). Chicago: Precedent Press.

Nussbaum, M. (2001). *Upheavals of thought: The intelligence of emotions*. New York: Cambridge University Press.

Nussbaum, M. (2004). *Hiding from humanity: Disgust, shame, and the law*. Princeton, NJ: Princeton University Press.

Olson, J. (2004). Buck-passing and the wrong kind of reasons. *Philosophical Quarterly, 54(215)*, 295–300.

O'Neill, O. (1998). Consistency in action. In J. Rachels (Ed.), *Ethical theory* (pp. 504–529). Oxford: Oxford University Press.

Parfit, D. (2001). Rationality and reasons. In D. Egonsson, J. Josefsson, B. Petersson, & T. Rønnow-Rasmussen (Eds.), *Exploring practical philosophy: From action to values* (pp. 17–39). Aldershot: Ashgate.

Pauer-Studer, H., & Velleman, J. D. (2011). Distortions of normativity. *Ethical Theory and Moral Practice, 14*(3), 329–356. doi:10.1007/s10677-010-9246-7.

Paxton, J. M., Bruni, T., & Greene, J. D. (2014). Are "counter-intuitive" deontological judgments really counter-intuitive? An empirical reply to Kahane et al. (2012). *Social Cognitive and Affective Neuroscience, 9*(9), 1368–1371. doi:10.1093/scan/nst102.

Paxton, J. M., & Greene, J. D. (2010). Moral reasoning: Hints and allegations. *Topics in Cognitive Science, 2*(3), 511–527. doi:10.1111/j.1756-8765.2010.01096.x.

Paxton, J. M., Ungar, L., & Greene, J. D. (2012). Reflection and reasoning in moral judgment. *Cognitive Science, 36*(1), 163–177. doi:10.1111/j.1551-6709.2011.01210.x.

Peirce, C. S. (1877). The fixation of belief. *Popular Science Monthly, 12*, 1–15.

Petrinovich, L., O'Neill, P., & Jorgensen, M. (1993). An empirical study of moral intuitions: Toward an evolutionary ethics. *Journal of Personality and Social Psychology, 64*(3), 467–478. doi:10.1037/0022-3514.64.3.467.

Pettit, P. (2001). *A theory of freedom: From the psychology to the politics of agency*. Oxford: Oxford University Press.

Piaget, J. (1965). *The moral judgment of the child*. New York: Free Press.

Pinillos, N., Smith, N., Nair, G. S., Marchetto, P., & Mun, C. (2011). Philosophy's new challenge: Experiments and intentional action. *Mind & Language, 26*(1), 115–139. doi:10.1111/j.1468-0017.2010.01412.x.

Pizarro, D. (2000). Nothing more than feelings? The role of emotions in moral judgment. *Journal for the Theory of Social Behaviour, 30*(4), 355–375. doi:10.1111/1468-5914.00135.

Pizarro, D. A., & Bloom, P. (2003). The intelligence of the moral intuitions: Comment on Haidt (2001). *Psychological Review, 110*(1), 193–196. doi:10.1037/0033-295X.110.1.193.

Plakias, A. (2013). The good and the gross. *Ethical Theory and Moral Practice, 16*(2), 261–278. doi:10.1007/s10677-012-9334-y.

Plessner, H. (1928). *Die Stufen des Organischen und der Mensch.* Berlin: Einleitung in die Philosophische Anthropologie.

Pohlman, H. (1999). *The whole truth: A case of murder on the Appalachian Trail.* Amherst, MA: University of Massachusetts Press.

Poldrack, R. A. (2006). Can cognitive processes be inferred from neuroimaging data? *Trends in Cognitive Sciences, 10*(2), 59–63. doi:10.1016/j.tics.2005.12.004.

Poldrack, R. A., & Wagner, A. D. (2004). What can neuroimaging tell us about the mind? Insights from prefrontal cortex. *Current Directions in Psychological Science, 13*(5), 177–181. doi:10.1111/j.0963-7214.2004.00302.x.

Pollard, B. (2003). Can virtuous actions be both habitual and rational? *Ethical Theory and Moral Practice, 6*(4), 411–425. doi:10.1023/B:ETTA.0000004561.34480.d4.

Pollard, B. (2005a). Naturalizing the space of reasons. *International Journal of Philosophical Studies, 13*(1), 69–82. doi:10.1080/0967255042000324344.

Pollard, B. (2005b). The rationality of habitual actions. *Proceedings of the Durham-Bergen Philosophy Conference, 1*, 39–50.

Pollock, J. (1986). A theory of moral reasoning. *Ethics, 96*(3), 506–523.

Pollock, J. (1987). Defeasible reasoning. *Cognitive Science, 11*(4), 481–518. doi:10.1207/s15516709cog1104_4.

Prichard, H. A. (1912). Does moral philosophy rest on a mistake? *Mind, 21*, 21–37.

Prinz, J. (2004). *Gut reactions: A perceptual theory of emotion.* New York: Oxford University Press.

Prinz, J. (2006). The emotional basis of moral judgment. *Philosophical Explorations, 9*(1), 29–43. doi:10.1080/13869790500492466.

Prinz, J. (2007). *The emotional construction of morals.* New York: Oxford University Press.

Prinz, J. (2008). Empirical philosophy and experimental philosophy. In J. Knobe & S. Nichols (Eds.), *Experimental philosophy* (pp. 189–208). New York: Oxford University Press.

Prinz, J. (2011a). Is empathy necessary for morality? In A. Coplan & P. Goldie (Eds.), *Empathy: Philosophical and psychological perspectives* (pp. 519–538). Oxford: Oxford University Press.

Prinz, J. (2011b). Against empathy. *Southern Journal of Philosophy, 49* (Suppl. 1), 214–233. doi:10.1111/j.2041-6962.2011.00069.x.

Prinz, J. (2015). Naturalizing metaethics. In T. Metzinger & J. M. Windt (Eds.), *Open Mind*. Frankfurt: MIND Group. doi:10.15502/9783958570511.

Prinz, J., & Nichols, S. (2010). Moral emotions. In J. Doris & the The Moral Psychology Research Group (Ed.), *The moral psychology handbook* (pp. 111–147). New York: Oxford University Press.

Pritchard, D. (2002). Two forms of epistemological contextualism. *Grazer Philosophische Studien, 64*, 19–55.

Pritchard, D. (2005). *Epistemic luck*. New York: Oxford University Press.

Pugmire, D. (2005). *Sound sentiments: Integrity in the emotions*. New York: Oxford University Press.

Rabinowicz, W., & Rønnow-Rasmussen, T. (2004). The strike of the demon: On fitting pro-attitudes and value. *Ethics, 114(3)*, 391–424.

Rai, T. S., & Fiske, A. P. (2011). Moral psychology is relationship regulation: Moral motives for unity, hierarchy, equality, and proportionality. *Psychological Review, 118(1)*, 57–75. doi:10.1037/a0021867.

Railton, P. (2014). The affective dog and its rational tale: Intuition and attunement. *Ethics, 124*(4), 813–859. doi:10.1086/675876.

Rawls, J. (1971). *A theory of justice*. Cambridge, MA: Harvard University Press.

Reed, D. C. (2008). A model of moral stages. *Journal of Moral Education, 37*(3), 357–376. doi:10.1080/03057240802227759.

Riedl, K., Jensen, K., Call, J., & Tomasello, M. (2012). No third-party punishment in chimpanzees. *Proceedings of the National Academy of Sciences of the United States of America, 109*(37), 14824–14829.

Rini, R. (2013). Making psychology normatively significant. *Journal of Ethics, 17*(3), 257–274. doi:10.1007/s10892-013-9145-y.

Ronson, J. (2011). *The psychopath test: A journey through the madness industry*. London: Picador.

Roskies, A. (2003). Are ethical judgments intrinsically motivational? Lessons from "acquired sociopathy." *Philosophical Psychology, 16*(1), 51–66. doi:10.1080/09515080 32000067743.

Royzman, E. B., Kim, K., & Leeman, R. F. (2015). The curious tale of Julie and Mark: Unraveling the moral dumbfounding effect. *Judgment and Decision Making, 10*(4), 296–313.

Rozin, P., & Fallon, A. (1987). A perspective on disgust. *Psychological Review, 94*(1), 23–41. doi:10.1037/0033-295X.94.1.23.

Rozin, P., Haidt, J., McCauley, C., Dunlop, L., & Ashmore, M. (1999). Individual differences in disgust sensitivity: Comparisons and evaluations of paper-and-pencil versus behavioral measures. *Journal of Research in Personality, 33*(3), 330–351. doi:10.1006/jrpe.1999.2251.

Rozin, P., Lowery, L., Imada, S., & Haidt, J. (1999). The CAD triad hypothesis: A mapping between three moral emotions (contempt, anger, disgust) and three moral codes (community, autonomy, divinity). *Journal of Personality and Social Psychology, 76*(4), 574–586. doi:10.1037/0022-3514.76.4.574.

Rozin, P., & Singh, L. (1999). The moralization of cigarette smoking in the United States. *Journal of Consumer Psychology, 8*(3), 321–337. doi:10.1207/s15327663jcp0803_07.

Rudman, L., Ashmore, R., & Gary, M. L. (2001). "Unlearning" automatic biases: The malleability of implicit prejudice and stereotypes. *Journal of Personality and Social Psychology, 81*(5), 856–868. doi:10.1037/0022-3514.81.5.856.

Ryle, G. (1949). *The concept of mind*. London: Penguin Books.

Saltzstein, H. D., & Kasachkoff, T. (2004). Haidt's moral intuitionist theory: A psychological and philosophical critique. *Review of General Psychology, 8*(4), 273–282. doi:10.1037/1089-2680.8.4.273.

Sauer, H. (2011a). Social intuitionism and the psychology of moral reasoning. *Philosophy Compass, 6*(10), 708–721. doi:10.1111/j.1747-9991.2011.00437.x.

Sauer, H. (2011b). The appropriateness of emotions: Moral judgment, moral emotions, and the conflation problem. *Ethical Perspectives, 18*(1), 107–140. doi:10.2143/EP.18.1.2066215.

Sauer, H. (2012a). Morally irrelevant factors: What's left of the dual-process model ofmoral cognition? *Philosophical Psychology, 25*(6), 783–811. doi:10.1080/09515089.2011.631997.

Sauer, H. (2012b). Educated intuitions: Automaticity and rationality in moral Judgment. *Philosophical Explorations, 15*(3), 255–275. doi:10.1080/13869795.2012.706822.

Sauer, H. (2012c). Psychopaths and filthy desks: Are emotions necessary and sufficient for moral judgment? *Ethical Theory and Moral Practice, 15*(1), 95–115. doi:10.1007/s10677-011-9274-y.

Sauer, H. (2015). Can't we all disagree more constructively? Moral foundations, moral reasoning, and political disagreement. *Neuroethics, 8*(2), 153–169. doi:10.1007/s12152-015-9235-6.

Saver, J. L., & Damasio, A. R. (1991). Preserved access and processing of social knowledge in a patient with acquired sociopathy due to ventromedial frontal damage. *Neuropsychologia, 29*(12), 1241–1249. doi:10.1016/0028-3932(91)90037-9.

Scanlon, T. M. (1998). *What we owe to each other*. Cambridge, MA: Harvard University Press.

Scarantino, A. (2010). Insights and blindspots of the cognitivist theory of emotions. *British Journal for the Philosophy of Science, 61*(4), 729–768. doi:10.1093/bjps/axq011.

Schaich Borg, J., Hynes, C., Van Horn, J., Grafton, S., & Sinnott-Armstrong, W. (2006). Consequences, action and intention as factors in moral judgments: An fMRI Investigation. *Journal of Cognitive Neuroscience, 18*(5), 803–817. doi:10.1162/jocn.2006.18.5.803.

Scheler, M. (2007). *Die Stellung des Menschen im Kosmos*. Bonn: Bouvier.

Schiller, F. (2004). *Über Anmut und Würde*. In F. Schiller, *Sämtliche Werke*, Vol. 5: *Erzählungen/Theoretische Schriften* (pp. 433–489). Munich: Hanser.

Schnall, S., Haidt, J., Clore, G. L., & Jordan, A. H. (2008). Disgust as embodied moral judgment. *Personality and Social Psychology Bulletin, 34*(8), 1096–1109. doi:10.1177/0146167208317771.

Schroeder, M. (2007). *Slaves of the passions*. Oxford: Oxford University Press.

Schroeder, M. (2008). *Being for: Evaluating the semantic program of expressivism*. New York: Oxford University Press.

Schwartz, N., & Clore, G. L. (1983). Mood, misattribution, and judgments of well-being: Informative and directive functions of affective states. *Journal of Personality and Social Psychology, 45*(3), 513–523. doi:10.1037/0022-3514.45.3.513.

Schwitzgebel, E., & Ellis, J. (forthcoming). Rationalization in moral and philosophical thought. In J. F. Bonnefon & B. Tremoliere (Eds.), *Moral inferences*. Psychology Press.

Shope, R. (1978). The conditional fallacy in contemporary philosophy. *Journal of Philosophy, 75*(8), 397–413. doi:10.2307/2025564.

Sie, M. (2009). Moral agency, conscious control, and deliberative awareness. *Inquiry, 52*(5), 516–531. doi:10.1080/00201740903302642.

Simon, H., & Newell, A. (1958). Heuristic problem solving: The next advance in operations research. *Operations Research, 6*(1), 1–10.

Singer, P. (1972). Famine, affluence, and morality. *Philosophy & Public Affairs, 1*(3), 229–243.

Singer, P. (2005). Ethics and intuitions. *Journal of Ethics, 9(3)*, 331–352. doi:10.1007/s10892-005-3508-y.

Sinnott-Armstrong, W. (1996). Moral skepticism and justification. In W. Sinnott-Armstrong and M. Timmons (Eds.), *Moral knowledge? New readings in moral epistemology* (pp. 3–48). New York: Oxford University Press.

Sinnott-Armstrong, W. (2006). *Moral skepticisms*. New York: Oxford University Press.

Sinnott-Armstrong, W., & Wheatley, T. (2012). The disunity of morality and why it matters to philosophy. *Monist, 95*(3), 355–377.

Sinnott-Armstrong, W., Young, L., & Cushman, F. (2010). Moral intuitions. In J. Doris, & the The Moral Psychology Research Group (Ed.), *The moral psychology handbook*. New York: Oxford University Press.

Slote, M. (2004). Moral sentimentalism. *Ethical Theory and Moral Practice, 7(1)*, 3–14. doi:10.1023/B:ETTA.0000019982.56628.c3.

Smetana, J. G. (1984). Toddlers' social interactions regarding moral and conventional transgressions. *Child Development, 55*(5), 1767–1776. doi:10.2307/1129924.

Smetana, J. G. (1989). Toddlers' social interactions in the context of moral and conventional transgressions in the home. *Developmental Psychology, 25*(4), 499–508. doi:10.1037//0012-1649.25.4.499.

Smith, M. (1994). *The moral problem*. Malden, MA: Blackwell.

Smith, M. (1995). Internal reasons. *Philosophy and Phenomenological Research, 55*(1), 109–131.

Sneddon, A. (2007). A social model of moral dumbfounding: Implications for studying moral reasoning and moral judgment. *Philosophical Psychology, 20*(6), 731–748. doi:10.1080/09515080701694110.

Sneddon, A. (2011). *Like-minded: Externalism and moral psychology*. Cambridge, MA: MIT Press.

Snow, N. E. (2006). Habitual virtuous actions and automaticity. *Ethical Theory and Moral Practice, 9*(5), 545–561. doi:10.1007/s10677-006-9035-5.

Snow, N. E. (2010). *Virtue as social intelligence: An empirically grounded theory*. New York: Routledge.

Solomon, R. C. (1976). *The passions: Emotions and the meaning of life*. Indianapolis, IN: Hackett.

Sosa, E. (1991). Knowledge and intellectual virtue. In E. Sosa, *Knowledge in perspective: Selected essays in epistemology* (pp. 225–244). Cambridge: Cambridge University Press.

Sosa, E. (1999). How must knowledge be modally related to what is known? *Philosophical Topics, 26*(1–2), 373–384. doi:10.5840/philtopics1999261/229.

Southwood, N. (2011). The moral/conventional distinction. *Mind, 120*(479), 761–802. doi:10.1093/mind/fzr048.

Southwood, N., & Eriksson, L. (2011). Norms and conventions. *Philosophical Explorations, 14*(2), 195–217. doi:10.1080/13869795.2011.569748.

Sripada, C. S., & Stich, S. (2006). A framework for the psychology of norms. In P. Carruthers, S. Laurence, & S. Stich (Eds.), *The innate mind: Culture and cognition*. Oxford University Press. doi:10.1093/acprof:oso/9780195310139.003.0017.

Stevenson, C. L. (1937). The emotive meaning of ethical terms. *Mind, 46*, 14–31.

Stevenson, C. L. (1966). Ethical fallibility. In R. T. DeGeorge (Ed.), *Ethics and society*. Garden City, NY: Doubleday.

Stocker, M., & Hegeman, E. (1996). *Valuing emotions*. Cambridge, MA: Cambridge University Press.

Strawson, P. F. (1962). Freedom and resentment. *Proceedings of the British Academy, 48*, 1–25.

Street, S. (2006). A Darwinian dilemma for realist theories of value. *Philosophical Studies, 127(1)*, 109–166. doi:10.1007/s11098-005-1726-6.

Strohminger, N. (2014). Disgust talked about. *Philosophy Compass, 9*(7), 478–493. doi:10.1111/phc3.12137.

Sunstein, C. R. (2005). Moral heuristics. *Behavioral and Brain Sciences, 28*(*4*), 531–573. doi:10.1017/S0140525X05000099.

Tappolet, C. (forthcoming). Emotions, perceptions, and emotional illusions. In C. Calabi (Ed.), *The crooked oar, the moon's size, and the Kanizsa triangle: Essays on perceptual illusions*. Cambridge, MA: MIT Press.

Thomas, A. (2006). *Value and context: The nature of moral and political knowledge*. Oxford: Oxford University Press.

Thomas, A. (2011). Another particularism: Reasons, status, and defaults. *Ethical Theory and Moral Practice, 14*(2), 151–167. doi:10.1007/s10677-010-9247-6.

Thompson, V., Prowse Turner, J., & Pennycook, G. (2011). Intuition, reason, and metacognition. *Cognitive Psychology, 63(3)*, 107–140. doi:10.1016/j.cogpsych.2011.06.001.

Thomson, J. J. (1976). Killing, letting die, and the trolley problem. *Monist, 59*(2), 204–217.

Tiberius, V. (2013). In defense of reflection. *Philosophical Issues, 23*(1), 223–243. doi:10.1111/phis.12011.

Timmons, M. (1996). Outline of a contextualist moral epistemology. In W. Sinnott-Armstrong and M. Timmons (Eds.), *Moral knowledge? New readings in moral epistemology* (pp. 293–325). New York: Oxford University Press.

Timmons, M. (1999). *Morality without foundations: A defense of ethical contextualism.* New York: Oxford University Press.

Toulmin, S. (1958). *The uses of argument.* Cambridge: Cambridge University Press.

Trouche, E., Johansson, P., Hall, L., & Mercier, H. (2015). The selective laziness of reasoning. *Cognitive Science.* doi:10.1111/cogs.12303.

Turiel, E. (1983). *The development of social knowledge: Morality and convention.* Cambridge: Cambridge University Press.

Tversky, A., & Kahneman, D. (1981). The framing of decisions and the psychology of choice. *Science, 211(4481)*, 453–458. doi:10.1126/science.7455683.

Tybur, J. M., Lieberman, D., Kurzban, R., & DeScioli, P. (2013). Disgust: Evolved function and structure. *Psychological Review, 120(1)*, 65–84. doi:10.1037/a0030778.

Uhlmann, E. L., & Cohen, G. L. (2005). Constructed criteria: Redefining merit to justify discrimination. *Psychological Science, 16*(6), 474–480.

Uhlmann, E. L., Pizarro, D. A., Tannenbaum, D., & Ditto, P. H. (2009). The motivated use of moral principles. *Judgment and Decision Making, 4*(6), 476–491.

Unger, P. (1996). *Living high and letting die: Our illusion of innocence.* New York: Oxford University Press.

Unwin, N. (1999). Quasi-realism, negation, and the Frege–Geach problem. *Philosophical Quarterly, 49*(196), 337–352. doi:10.1111/1467-9213.00146.

Unwin, N. (2001). Norms and negation: A problem for Gibbard's logic. *Philosophical Quarterly, 51*(202), 60–75. doi:10.1111/1467-9213.00214.

Valdesolo, P., & DeSteno, D. (2006). Manipulations of emotional context shape moral judgment. *Psychological Science, 17*(6), 476–477. doi:10.1111/j.1467-9280.2006.01731.x.

van Roojen, M. (1996). Expressivism and irrationality. *Philosophical Review, 105*(3), 311–335.

van Roojen, M. (2000). Motivational internalism: A somewhat less idealized account. *Philosophical Quarterly, 50*(199), 233–241. doi:10.1111/1467-9213.00183.

Västfjäll, D., Slovic, P., & Mayorga, M. (2015). Pseudoinefficacy: Negative feelings from children who cannot be helped reduce warm glow for children who can be helped. *Frontiers in Psychology, 6*(616). doi:10.3389/fpsyg.2015.00616.

Västfjäll, D., Slovic, P., Mayorga, M., & Peters, E. (2014). Compassion fade: Affect and charity are greatest for a single child in need. *PLoS One, 9*(6), e100115. doi:10.1371/journal.pone.0100115.

Velleman, J. D. (2000). *The possibility of practical reason.* New York: Oxford University Press.

Velleman, J. D. (2010). There are no "reasons for acting." Unpublished manuscript.

Vogt, B. A., & Absher, J. R. (2000). Human retrosplenial cortex: Where is it and is it involved in emotion? *Trends in Neurosciences, 23*(5), 195–196. doi:10.1016/S0166-2236(00)01579-4.

Wager, T. D., & Smith, E. E. (2003). Neuroimaging studies of working memory: A meta-analysis. *Cognitive, Affective & Behavioral Neuroscience, 3*(4), 255–274. doi:10.3758/CABN.3.4.255.

Wallace, R. J. (1994). *Responsibility and the moral sentiments.* Cambridge, MA: Harvard University Press.

Way, J. (2012). Transmission and the wrong kind of reason. *Ethics, 122*(3), 489–515. doi:10.1086/664749.

Weber, M. (1988). Die Protestantische Ethik und der Geist des Kapitalismus. In *Gesammelte Aufsätze zur Religionssoziologie I* (pp. 1–206). Tübingen: Mohr.

Wegner, D. (2002). *The illusion of conscious will.* Cambridge, MA: MIT Press.

Wellmann, C. (1971). *Challenge and response: Justification in ethics.* Carbondale: Southern Illinois University Press.

Wheatley, T., & Haidt, J. (2005). Hypnotic disgust makes moral judgment more severe. *Psychological Science, 16*(10), 780–784. doi:10.1111/j.1467-9280.2005.01614.x.

Wielenberg, E. J. (2014). *Robust ethics: The metaphysics and epistemology of godless normative realism.* Oxford: Oxford University Press.

Wiggins, D. (1998). A sensible subjectivism? In *Needs, values, and truth* (pp. 185–214). Oxford: Clarendon Press.

Willaschek, M. (2002). Moralisches Urteil und begründeter Zweifel: Eine kontextualistische Konzeption der Rechtfertigung moralischer Urteile. In A. Beckermann & C. Nimtz (Eds.), *Argument und Analyse: Ausgewählte Sektionsvorträge des 4. internationalen Kongresses der Gesellschaft für Analytische Philosophie*, 630–641. Paderborn: Mentis.

Willaschek, M. (2007). Contextualism about knowledge and justification by default. *Grazer Philosophische Studien, 74(1)*, 251–272. doi:10.1163/9789401204651_014.

Williams, B. (1970). Deciding to believe. In H. Kiefer & M. Munitz (Eds.), *Language, belief, and metaphysics* (pp. 95–111). Albany, NY: SUNY Press.

Williams, B. (1981). Persons, character, and morality. In *Moral Luck* (pp. 1–19). Cambridge: Cambridge University Press.

Williams, M. (1988). Epistemological realism and the basis of scepticism. *Mind, 97*(387), 415–439. doi:10.1093/mind/XCVII.387.415.

Williams, M. (1999). *Groundless belief: An essay on the possibility of epistemology.* Princeton, NJ: Princeton University Press.

Williams, M. (2001). *Problems of knowledge: A critical introduction to epistemology.* New York: Oxford University Press.

Williams, M. (2008). Responsibility and reliability. *Philosophical Papers, 37*(1), 1–26. doi:10.1080/05568640809485212.

Wilson, T. D. (2002). *Strangers to ourselves: Discovering the adaptive unconscious.* Cambridge, MA: Belknap Press of Harvard University Press.

Wilson, T. D., & Brekke, N. (1994). Mental contamination and mental correction: Unwanted influences on judgments and evaluations. *Psychological Bulletin, 116(1)*, 117–142. doi:10.1037/0033-2909.116.1.117.

Wilson, T. D., Dunn, D. S., Kraft, D., & Lisle, D. J. (1989). Introspection, attitude change, and attitude-behavior consistency: The disruptive effects of explaining why we feel the way we do. In L. Berkowitz (Ed.), *Advances in experimental social psychology* (pp. 287–343). San Diego: Academic Press.

Wilson, T. D., & Schooler, J. W. (1991). Thinking too much: Introspection can reduce the quality of preferences and decisions. *Journal of Personality and Social Psychology, 60*(2), 181–192. doi:10.1037/0022-3514.60.2.181.

Winch, P. (1958). *The idea of a social science and its relation to philosophy.* London: Routledge & Kegan Paul.

Wittgenstein, L. (1969). *On certainty.* Oxford: Blackwell.

Wood, W., & Neal, D. T. (2007). A new look at habits and the habit-goal interface. *Psychological Review, 114*(4), 843–863. doi:10.1037/0033-295X.114.4.843.

Wood, W., Quinn, J. M., & Kashy, D. A. (2002). Habits in everyday life: Thought, emotion, and action. *Journal of Personality and Social Psychology, 83*(6), 1281–1297. doi:10.1037/0022-3514.83.6.1281.

Wynne, C. D. L. (2004). Animal behavior: Fair refusal by capuchin monkeys. *Nature, 428(6979*), 140. doi:10.1038/428140a.

Zagzebski, L. (1996). *Virtues of the mind: An inquiry into the nature of virtue and the ethical foundations of knowledge*. Cambridge: Cambridge University Press.

Name Index

Subject Index

Printed by Libri Plureos GmbH in Hamburg,
Germany

9 780262 546706